**PMP**
Project Management
Professional Strategy

GUIDE
PMBOK

SIXIEME EDITION

# 1. INTRODUCTION
## PROJECT MANAGEMENT

Presented by :
**Abdulrahman Al Merei**
PMO Director, ISO 21500 Civil Engineer
Certified Project Managers Trainer

## What is a Project

**Project** is a temporary endeavor undertaken to create a unique product, service, or result

The end of the project is reached when
- The project's objectives have been achieved;
- The objectives will not or cannot be met;
- Funding is exhausted
- The need for the project no longer
- The project is terminated for legal cause or convenience

## Fundamental elements of Project

Projects drive change the project moving an organization from one state to another state, **By** achieve a specific objective,

Projects enable business value creation **By** create benefits to Benefits may be tangible or intangible or both

| Tangible | Intangible |
|---|---|
| Monetary assets | Goodwill |
| Stockholder equity | Brand recognition |
| Utility | Public benefit |

### Project Initiation Context.

- Meet regulatory, legal, or social requirements;
- Satisfy stakeholder requests or needs;
- Implement or change business or technological strategies; and
- Create, improve, or fix products, processes, or services

## Importance of Project management

**Project management**
is the application of knowledge, skills, tools, and techniques to project activities to meet the project requirements.

 **Project Management** enables organizations to execute projects effectively and efficiently

Effective and efficient project management enables organizations to:
- Tie project results to business goals,
- Compete more effectively in their markets,
- Sustain the organization, and
- Respond to the impact of business environment changes on projects by appropriately adjusting project management plans

## Relation between Project, Program, Portfolio, & Operations

**Project**
May be managed as a stand-alone project, within a program, or within a portfolio.

**Program**
Group of related projects, subsidiary programs, and program activities managed in a coordinated manner to obtain benefits not available from managing them individually

**Portfolio**
Is a projects, programs, subsidiary portfolios, and operations managed as a group to achieve strategic objectives.

**Operations management**
Concerned with ongoing production of goods and/or services.

- **Program and project management focus on doing programs and projects the "right" way;**
- **Portfolio management focuses on doing the "right" programs and projects.**
- **Operations management ensures that business operations counties efficiently by using the optimal resources to transform input to output**

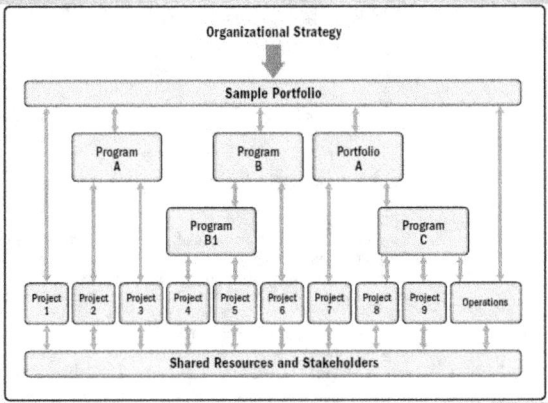

## Difference between Project, Program, Portfolio

| | Project | Program | Portfolio |
|---|---|---|---|
| **Definition** | • temporary endeavor<br>• to create a unique product, service, or result. | • group of related projects, sub programs, and activities<br>• to obtain benefits not available from managing them individually. | • collection of projects, programs, sub portfolios, and operations<br>• to achieve strategic objectives. |
| **Scope** | • defined objectives.<br>• Scope is progressively elaborated | • Encompasses the scopes of its components.<br>• the outputs of program components are delivered in a complementary manner. | • Organizational scope<br>• changes with the strategic objectives of the organization. |
| **Change** | Change managed and controlled. | accepts and adapts to change | Continuously monitor changes in the internal & external environments. |
| **Planning** | Progressively elaborate high-level information into detailed plans throughout the project life cycle. | high-level plans that track the interdependencies and progress of program components | maintain necessary processes and communication relative to the Aggregate portfolio. |
| **Management** | to meet the project objectives | coordinating the activities within program to ensure benefits are delivered as expected, | manage or coordinate portfolio management staff |
| **Monitoring** | monitor and control the work of producing the products, services, or results | Monitor the progress of program components to ensure the overall goals, schedules, budget, and benefits of the program will be met. | monitor strategic changes and aggregate resource allocation, performance results, and risk of the portfolio. |
| **Success** | measured by product and project quality, timeliness, budget compliance, and degree of customer satisfaction. | measured by ability to deliver its intended benefits and efficiency and effectiveness in delivering those benefits. | measured in terms of the aggregate investment performance and benefit realization of the portfolio. |

## Relation between Project, Program, Portfolio,& Operations

**ORGANIZATIONAL PROJECT MANAGEMENT (OPM) AND STRATEGIES**
- Portfolio management aligns portfolios with organizational strategies by selecting the right programs or projects, prioritizing the work, and providing the needed resources.
- Program management harmonizes its program components and controls interdependencies in order to realizes pecified benefits.
- Project management enables the achievement of organizational goals and objectives.

**Organizational project management (OPM).**
Defined as a framework in which portfolio, program, and project management are integrated with organizational enablers in order to achieve strategic objectives.

- Ensure that the organization undertakes the right projects
- Allocates critical resources appropriately.
- Ensure that all levels in the organization understand the strategic vision, the initiatives that support the vision, the objectives, and the deliverables.

## Relation between Project, Program, Portfolio, & Operations

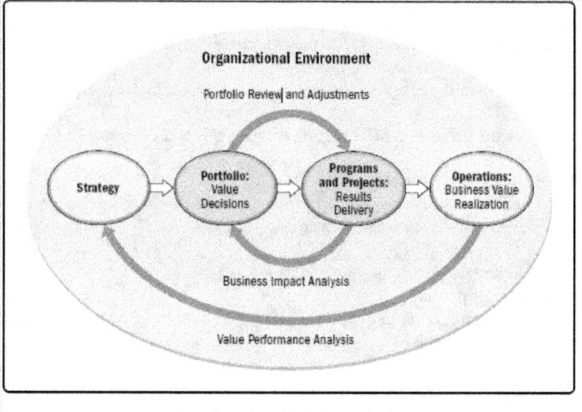

Figure 1-4. Organizational Project Management

## Project Lifecycle

**Project life cycle**
**Is the series of phases that a project passes through from its start to its completion.**

**It provides the basic framework for managing the project.**

Project life cycles **can be predictive or adaptive to accomplish the product.**

It is up to the project management team to determine the best life cycle for each project.

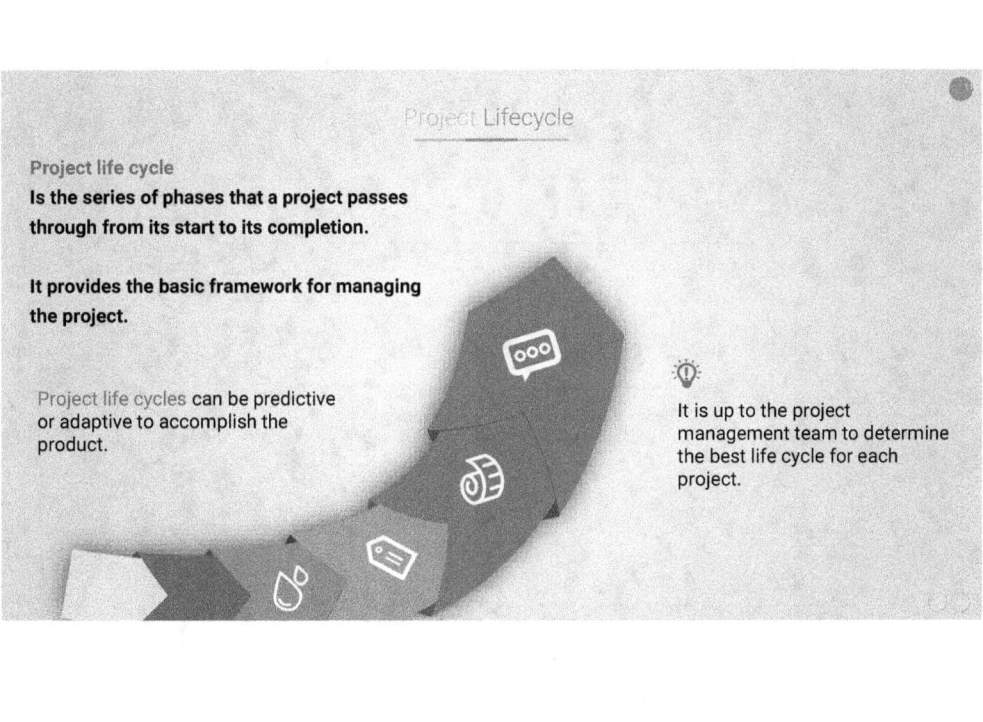

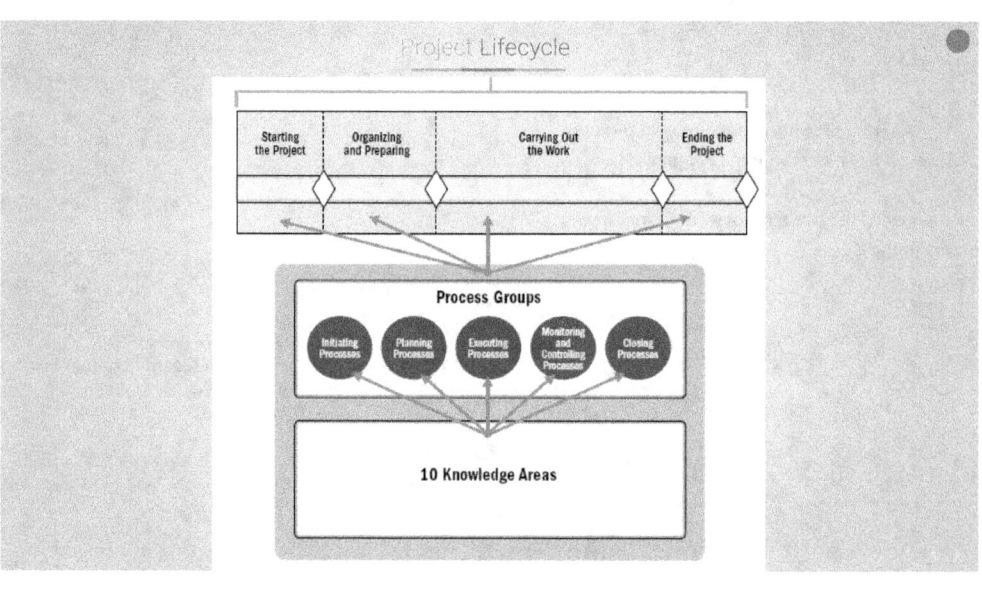

## Development Lifecycle

💡 **Development life cycles** one or more phases that are associated with the development of the product, service, or result.
can be predictive, iterative, incremental, adaptive, or a hybrid model:

- **Predictive life cycle** (waterfall) scope, time, and cost are determined in the early phases. Any changes to the scope are carefully managed..
- **Iterative life cycle,** the project scope is generally determined early, but time and cost estimates are routinely modified
Iterations develop the product through a series of repeated cycles
- **Incremental life cycle,** the deliverable is produced through a series of iterations that successively add functionality within a predetermined time frame.
- **Adaptive life cycles** are agile or change-driven life cycles, iterative, or incremental. The detailed scope is defined and approved before the start of an iteration.
- **hybrid life cycle** is a combination of a predictive and an adaptive life cycle.

## Development Lifecycle

### PHASE

**is a collection of logically related project activities described by attributes (Name, number, Duration, Resource requirements, etc.)**

Examples of phase names include but are not limited to: (Concept development, Feasibility study, Customer requirements, Solution development, Design, Prototype… etc.

### PHASE GATE

A phase gate, is held at the end of a phase. The project's performance and progress are compared to project and business documents (business case, Project charter, Project management plan ,Benefits management plan).

may be called (phase review, stage gate, kill point, and phase entrance or phase exit)

A decision (e.g., go/no-go decision) is made Depending on the organization, industry, or type of work.

## Project Management Processes

### PROJECT MANAGEMENT PROCESSES

- Every project management process produces one or more outputs from one or more inputs by
- using appropriate project management tools and techniques
- Project management processes are logically linked by the outputs they produce.
- Processes may contain overlapping activities that occur throughout the project.
- The number of process iterations and interactions between processes varies based on the needs of the project.

**Processes generally fall into one of three categories:**
- Processes used once or at predefined points in the project.
- Processes that are performed periodically as needed
- Processes that are performed continuously throughout the project.

# Project Management Processes

 **Project Management Process Group**

is a logical grouping of project management processes to achieve specific project objectives.
- Initiating Process Group.
- Planning Process Group.
- Executing Process Group.
- Monitoring and Controlling Process Group.
- Closing Process Group.

**Project management Knowledge Areas.**
1. Project Integration Management
2. Project Scope Management.
3. Project Schedule Management
4. Project Cost Management.
5. Project Quality Management
6. Project Resource Management.
7. Project Communications Management.
8. Project Risk Management
9. Project Procurement Management.
10. Project Stakeholder Management..

| Knowledge Areas | Project Management Process Groups | | | | |
| --- | --- | --- | --- | --- | --- |
| | Initiating | Planning | Executing | Monitoring and Controlling | Closing |
| Project Integration Management | 4.1 Develop Project Charter | 4.2 Develop Project Management Plan | 4.3 Direct and Manage Project Work<br>4.4 Manage Project Knowledge | 4.5 Monitor and Control Project Work<br>4.6 Perform Integrated Change Control | 4.7 Close Project |
| Project Scope Management | | 5.1 Plan Scope Management<br>5.2 Collect Requirements<br>5.3 Define Scope<br>5.4 Create WBS<br>5.5 Validate Scope | | 5.6 Control Scope | |
| Project Schedule Management | | 6.1 Plan Schedule<br>6.2 Define Activities<br>6.3 Sequence Activities<br>6.4 Estimate Activity Durations<br>6.5 Develop Schedule Management | | 6.6 Control Schedule | |
| Project Cost Management | | 7.1 Plan Cost Management<br>7.2 Estimate Costs<br>7.3 Determine Budge | | 7.4 Control Costs | |
| Project Quality Management | | 8.1 Plan Quality Management | 8.2 Manage Quality | 8.3 Control Quality | |
| Project Resource Management | | 9.1 Plan Resource Management<br>9.2 Estimate Activity Resources | 9.4 Acquire Resources<br>9.5 Develop Team<br>9.6 Manage Team | 9.7 Control Resources | |
| Project Communications Management | | 10.1 Plan Communications Management | 10.2 Manage Communications | 10.3 Monitor Communications | |
| Project Risk Management | | 11.1 Plan Risk Management<br>11.2 Identify Risks<br>11.3 Perform Qualitative Risk Analysis<br>11.4 Perform Quantitative Risk Analysis<br>11.5 Plan Risk Responses | 11.6 Implement Risk Responses | 11.7 Monitor Risks | |
| Project Procurement Management | | 12.1 Plan Procurement Management | 12.2 Conduct Procurements | 12.3 Control Procurements | |
| Project Stakeholder Management | 13.1 Identify Stakeholders | 13.2 Plan Stakeholder Engagement | 13.4 Manage Stakeholder Engagement | 13.5 Monitor Stakeholder Engagement | |

## Project Management Processes

### Project Management Data and Information

Project data are regularly collected and analyzed throughout the project life cycle

- **Work performance data.** The raw observations and measurements identified during activities performed to carry out the project work.

- **Work performance information.** The performance data collected from various controlling processes, analyzed in context and integrated based on relationships across areas

- **Work performance reports.** The physical or electronic representation of work performance information compiled in project documents, which is intended to generate decisions or raise issues, actions, or awareness.

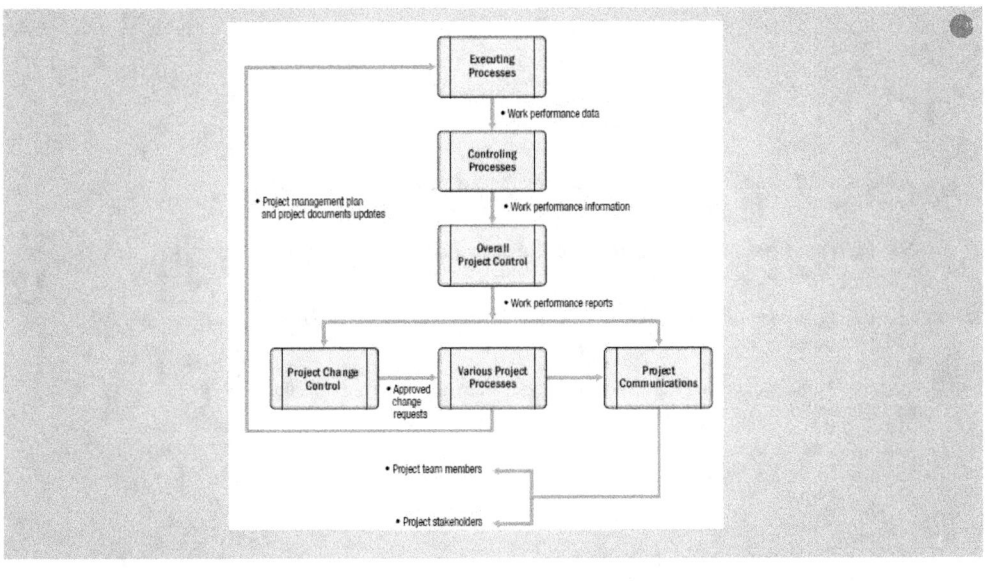

## Project Management Processes

**TAILORING**

Is a selection of The appropriate project management processes, inputs, tools, techniques, outputs, and life cycle phases should be selected to manage a project.

- Tailoring is necessary because each project is unique; not every process, tool, technique, input, or output identified.

- Tailoring should address the competing constraints of scope, schedule, cost, resources, quality, and risk.

- The project manager collaborates with the project team, sponsor, organizational management, or some combination thereof,

Project Management Business Documents

## Project Business Documents

- **Project business case:** A documented economic feasibility study used to establish the validity of the benefits of a selected component lacking sufficient definition and that is used as a basis for the authorization of further project management activities.

- **Project benefits management plan:** The documented explanation defining the processes for creating, maximizing, and sustaining the benefits provided by a project.

## Interrelationship of Needs Assessment and Critical Business/Project Documents

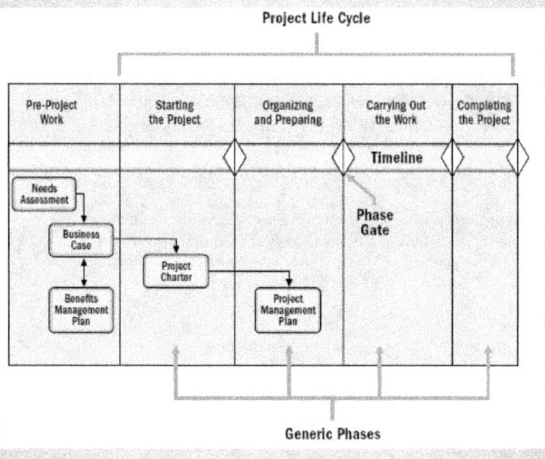

## Project Management Business Documents

### PROJECT BUSINESS CASE
The project business case is a documented economic feasibility study used to establish the validity of the benefits of a selected component lacking sufficient definition
- used as a basis for the authorization of further project management activities.
- It helps measure the project success at the end of the project against the project objectives.

A business case may include but is not limited to documenting the following:
- Business needs:
- Determination of what is prompting the need for action;
- Situational statement documenting the business problem or opportunity to be addressed including the value to be delivered to the organization;
  - Identification of stakeholders affected; and
  - Identification of the scope.
  - Analysis of the situation:
  - Identification of organizational strategies, goals, and objectives;
  - Identification of root cause(s) of the problem or main contributors of an opportunity;
  - Gap analysis of capabilities needed for the project versus existing capabilities of the organization;
  - Identification of known risks;
  - Identification of critical success factors;
  - Identification of decision criteria by which the various courses of action may be assessed;

## Project Management Business Documents

### PROJECT BENEFITS MANAGEMENT PLAN

Is the document that describes how and when the benefits of the project will be delivered, and describes the mechanisms that should be in place to measure those benefits.

A project benefit is defined as an outcome of actions, behaviors, products, services, or results that provide value to the sponsoring

benefits and may include but is not limited to documenting the following:
- Target benefits
- Strategic
- Timeframe for realizing benefits
- Benefits owner
- Metrics
- Assumptions
- Risks

## Project Management Business Documents

### PROJECT CHARTER AND PROJECT MANAGEMENT PLAN

**The project charter** is defined as a document issued by the project sponsor that formally authorizes the existence of a project and provides the project manager with the authority to apply organizational resources to project activities.

**The project management plan** is defined as the document that describes how the project will be executed, monitored, and controlled.

## Project Management Business Documents

**PROJECT SUCCESS MEASURES**
- Completing the project benefits management plan;
- Meeting the agreed-upon financial measures documented in the business case.
  - Net present value (NPV),
  - Return on investment (ROI),
  - Internal rate of return (IRR),
  - Payback period (PBP), and
  - Benefit-cost ratio (BCR).
- Meeting business case nonfinancial objectives;
- Completing movement of an organization from its current state to the desired state;
- Fulfilling contract terms and conditions;
- Meeting organizational strategy, goals, and objectives;
- Achieving stakeholder satisfaction;
- Acceptable customer/end-user adoption;
- Integration of deliverables into the organization's operating environment;
- Achieving agreed-upon quality of delivery;
- Meeting governance criteria; and
- Achieving other agreed-upon success measures or criteria (e.g., process throughput).

What   Why   When   Where   How   who

## 2. THE ENVIRONMENT
IN WHICH PROJECT OPERATE

Presented by :
**Nasser Al Mohimeed**
PMO Director, ISO 21500 Lead Project Manager
Certified Project Managers Trainer

## Project Environment

### Enterprise environmental factors
Refer to conditions, not under the control of the project team, that influence, constrain, or direct the project positive or negative,.

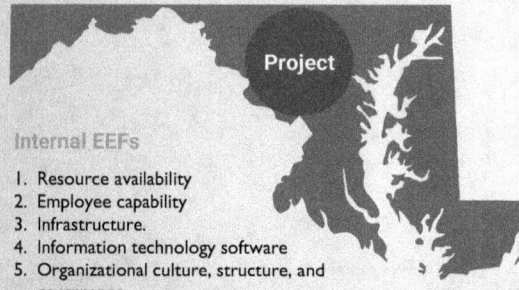

**Project**

### Internal EEFs
1. Resource availability
2. Employee capability
3. Infrastructure.
4. Information technology software
5. Organizational culture, structure, and governance.
6. Geographic distribution of facilities and resources.

### External EEFs
1. Marketplace conditions
2. Social and cultural influences and issues
3. Legal restrictions
4. Commercial databases
5. risk study information, and risk databases.
6. Academic research
7. Government or industry standards
8. Financial considerations
9. Physical environmental elements.

## Project Environment

### Organizational process assets

**OPAs** plans, processes, policies, procedures, and knowledge bases specific to and used by the performing organization.

- **Processes, policies, and procedures**
- **Organizational knowledge bases**

## Project Environment

**OPA- Processes, policies, and procedures**

**Related to Initiating and Planning:**
- Guidelines and criteria for tailoring processes and procedures to satisfy project needs
- organizational standards
- Product and project life cycles, and methods and procedures
- Templates
- Preapproved supplier lists

**Executing, Monitoring, and Controlling:**
- Change control procedures,
- Traceability matrices;
- Financial controls procedures
- Issue and defect management procedures
- Resource availability
- Organizational communication requirements
- Templates
- Product, service, or result verification and validation procedures.

**Related to Closing**
Project closure guidelines or requirements

## Project Environment

**OPA- Organizational knowledge bases**
- Configuration management knowledge repositories
- Financial data repositories
- Historical information and lessons learned knowledge repositories
- Issue and defect management data
- Data repositories for metrics and measurement
- Project files from previous projects

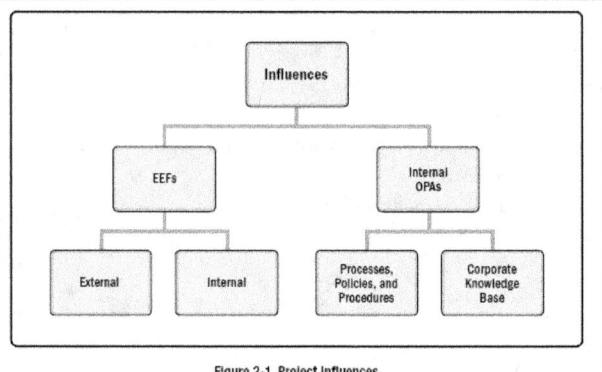

Figure 2-1. Project Influences

## Organizational System

PM needs to understand where responsibility, accountability, and authority reside within the organization.

✓ SUCCESS PROJECT

to effectively use his power, influence, competence, leadership, and political capabilities to successfully complete the project

A system is a collection of various components that together can produce results not obtainable by the individual components alone.

A component is an identifiable element within the project or organization that provides a particular function or group of related functions.

The interaction of the various system components creates the organizational culture and capabilities.

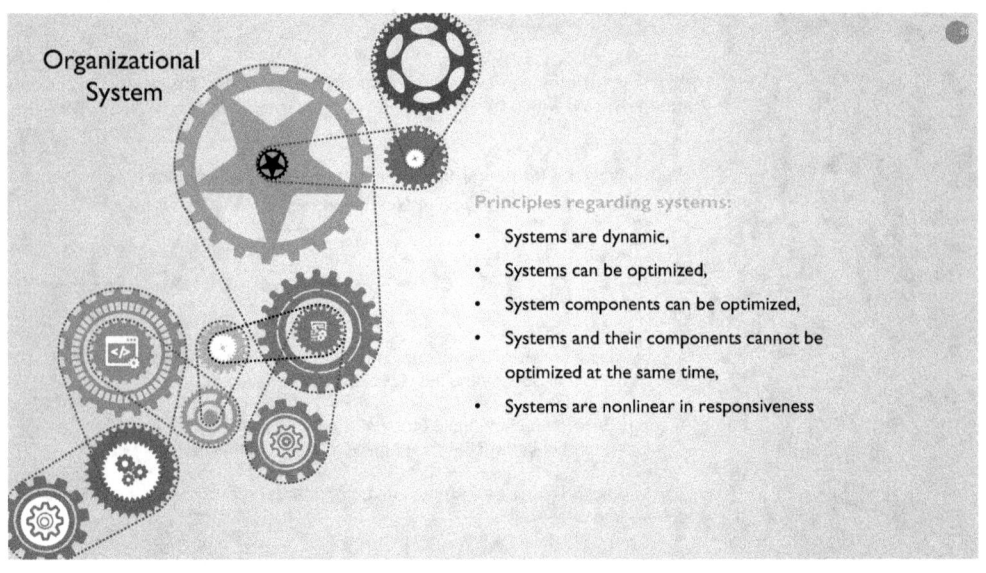

## Organizational System

**ORGANIZATIONAL GOVERNANCE**

Refers to organizational or structural arrangements at all levels of an organization designed to determine and influence the behavior of the organization's members
- Includes consideration of people, roles, structures, and policies;
- Requires providing direction and oversight through data and feedback.

**Governance framework**

includes but is not limited to:
- Rules,
- Policies,
- Procedures,
- Norms,
- Relationships,
- Systems, and
- Processes.

This framework influences how:
- Objectives of the organization are set and achieved,
- Risk is monitored and assessed, and
- Performance is optimized.

## Organizational System

**Governance of Portfolios, Programs, and Projects:**
Project governance refers to the framework, functions, and processes that guide project management activities in order to create a unique product, service, or result to meet organizational, strategic, and operational goals.

A governance framework should be tailored to the organizational culture, types of projects, and the needs of the organization in order to be effective.

## Organizational System

**MANAGEMENT ELEMENTS**
Are the components that comprise the key functions or principles of general management in the organization.

framework and the organizational structure type selected.
- Division of work using specialized skills and availability to perform work;
- Authority given to perform work;
- Responsibility to perform work appropriately assigned based on such attributes as skill and experience;
- Discipline of action (e.g., respect for authority, people, and rules);
- Unity of command (e.g., only one person gives orders for any action or activity to an individual);
- Unity of direction (e.g., one plan and one head for a group of activities with the same objective);
- General goals of the organization take precedence over individual goals;
- Paid fairly for work performed;
- Optimal use of resources;
- Clear communication channels;
- Right materials to the right person for the right job at the right time;
- Fair and equal treatment of people in the workplace;
- Clear security of work positions;
- Safety of people in the workplace;
- Open contribution to planning and execution by each person; and
- Optimal morale.

## Organizational Structure type

| Organizational Structure Type | Work Groups Arranged by: | PM Authority | PM Role | Resource Availability | Who Manages Budget? | PM |
|---|---|---|---|---|---|---|
| Organic or Simple | Flexible; people working side-by-side | Little or none | Part-time; may or may not be a designated job role like coordinator | Little or none | Owner or operator | Little or none |
| Functional | Job being done (e.g., engineering, manufacturing) | Little or none | Part-time; may or may not be a designated job role like coordinator | Little or none | Functional manager | Part-time |
| Multi-divisional | One of: product; production processes; portfolio; program; geographic region; customer type | Little or none | Part-time; may or may not be a designated job role like coordinator | Little or none | Functional manager | Part-time |
| Matrix – strong | By job function, with project manager as a function | Moderate to high | Full-time designated job role | Moderate to high | Project manager | Full-time |
| Matrix – weak | Job function | Low | Part-time; done as part of another job and not a designated job role like coordinator | Low | Functional manager | Part-time |
| Matrix – balanced | Job function | Low to moderate | Part-time; embedded in the functions as a skill and may not be a designated job role like coordinator | Low to moderate | Mixed | Part-time |
| Project-oriented | Project | High to almost total | Full-time designated job role | High to almost total | Project manager | Full-time |
| Virtual | Network structure with nodes at points of contact with other people | Low to moderate | Full-time or part-time | Low to moderate | Mixed | full-time/part-time |
| Hybrid | Mix of other types | Mixed | Mixed | Mixed | Mixed | Mixed |
| PMO | Mix of other types | High to almost total | Full-time designated job role | High to almost total | Project manager | Full-time |

## Project management office

**PMO** is an organizational structure that standardizes the project-related governance processes and facilitates the sharing of resources, methodologies, tools, and techniques

### Supportive
- Provide a consultative role by supplying templates, best practices, training, access to information, and lessons learned from other projects.
- serves as a project repository.
- The degree of is low.

### Controlling
- Provide support and require compliance
- The degree of control is moderate.

### Directive
- Directly managing the projects.
- Project managers are assigned by and report to the PMO.
- The degree of control is high.

## Project management office PMO

The PMO may:
- Make recommendations,
- Lead knowledge transfer,
- Terminate projects, and
- Take other actions, as required.

A primary function of a PMO is to support project managers
- Managing shared resources across all projects administered by the PMO;
- Identifying and developing project management methodology, best practices, and standards;
- Coaching, mentoring, training, and oversight;
- Monitoring compliance with project management standards, policies, procedures, and templates by means of
- project audits;
- Developing and managing project policies, procedures, templates, and other shared documentation (organizational process assets); and
- Coordinating communication across projects.

# 3. THE ROLE OF
## PROJECT MANAGER

Presented by :
**Nasser Al Mohimeed**
PMO Director, ISO 21500 Lead Project Manager
Certified Project Managers Trainer

# Project Manager

**Project manager** is the person assigned by the performing organization to lead the team responsible for achieving the project objectives

- **Membership and roles.** A large project may have more than 100 project members led by a project manager. Team
members may fulfill many different roles, such as design, manufacturing, and facilities management. The project members make up each leader's team.
- **Responsibility for team.** The project manager and conductor are both responsible for what their teams Produce
- **Knowledge and skills:** The project manager is not expected to perform every role on the project, but should possess project management knowledge, technical knowledge, understanding, and experience. The project manager provides the project team with leadership, planning, and coordination through communications.

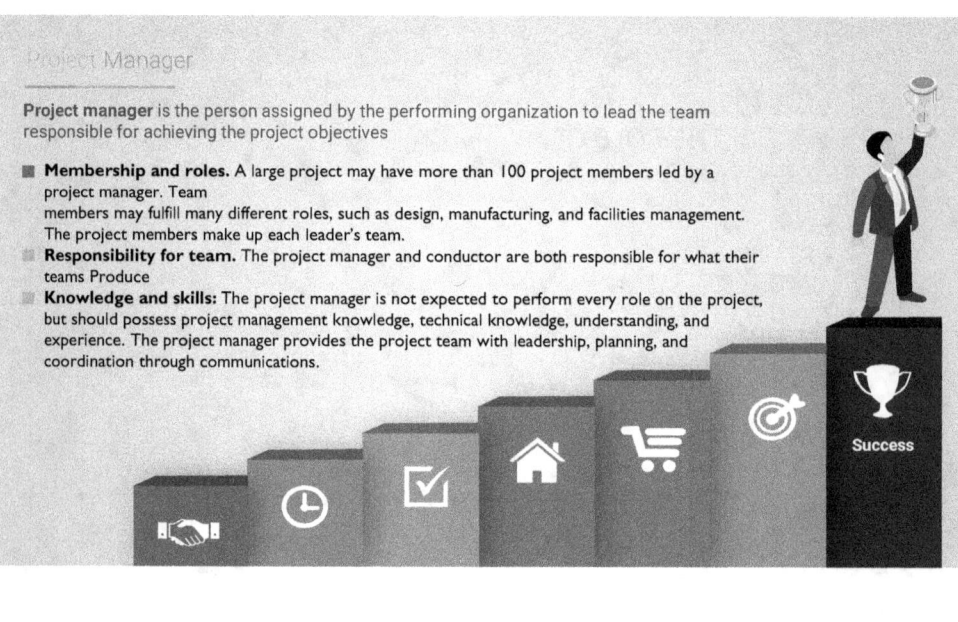

## The Project Manager's sphere of influence

- Project Manager fulfill numerous roles within their sphere of influence.

Circles (outer to inner):
- Stakeholders, Suppliers, Customers, End Users
- Sponsors, Governing Bodies, Steering Committees, PMOs
- Project Team, PPP Managers, Resource Managers
- Project Manager

**Project**
- Leads the project team to meet the project's objectives and stakeholders' expectations
- Performs communication roles between the project sponsor, team members, and other Stakeholder

**Organization**
- Interacts with other PMs within the same program
- Interacting with other PMs helps to create a positive influence for fulfilling project needs
- works with the project sponsor to address internal political and strategic issues that may impact the project

**Industry**
- stays informed about current industry trends (Technology, market niches, standers, economic force, etc.)

**Professional Discipline** Counting knowledge transfer and integration

**Across Disciplines** orient and educate other professionals regarding the value of a project management approach to the organization

## The Project Manager Competency

PMI studies applied the Project Manager Competency Development (PMCD) Framework to the skills needed by project managers through the use of The PMI Talent Triangle

**Technical project management.**
The knowledge, skills, and behaviors related to specific domains of project, program, and portfolio management.,

**Strategic and business management.**
The knowledge of and expertise in the industry and organization that enhanced performance and better delivers business outcomes

**Leadership**
The knowledge, skills, and behaviors needed to guide, motivate, and direct a team, to help an organization achieve its business goals

## The Project Manager Competency

 **Technical project management.**

The skills to effectively apply project management knowledge to deliver the desired outcomes for programs or projects.

Key skills including :

- ❖ Focus on technical project management elements for each project they manage.
  - Critical success factors for the project,
  - Schedule,
  - Selected financial reports,
  - Issue log.
- ❖ Tailor both traditional and agile tools, techniques, and methods for each project.
- ❖ Make time to plan thoroughly and prioritize diligently.
- ❖ Manage project elements, including, but not limited to, schedule, cost, resources, and risks.

## The Project Manager Competency

**Strategic and business management.**

Skills involve the ability to see the high-level overview of the organization and effectively negotiate and implement decisions and actions that support strategic alignment and innovation.

- ❖ Working knowledge of other functions such as finance, marketing, and operations.
- ❖ Domain knowledge (knowledgeable enough about the business) to be able to:
    - Explain to others the essential business aspects of a project;
    - Work with the project sponsor, team, and subject matter experts to develop an appropriate project delivery strategy;
    - Implement that strategy in a way that maximizes the business value of the project.
- ❖ the PM should be knowledgeable enough to explain to others the following aspects of the organization:
    - Strategy;
    - Mission;
    - Goals and objectives;
    - Products and services;
    - Operations (e.g., location, type, technology);
    - The market and the market condition, such as customers, state of the market.
    - Competition (e.g., what, who, position in the market place).

## The Project Manager Competency

 **Strategic and business management.**(continue)

❖ The PM should apply the following knowledge about the organization to the project to ensure alignment:
  - Strategy,
  - Mission,
  - Goals and objectives,
  - Priority,
  - Tactics, and
  - Products or services (e.g., deliverables).

❖ business and strategic factors Helps PM to determine which business factors could affect the project
  - Risks and issues,
  - Financial implications,
  - Cost versus benefits analysis
  - Business value,
  - Benefits realization expectations and strategies, and
  - Scope, budget, schedule, and quality.

## The Project Manager Competency

### LEADERSHIP
Leadership skills involve the ability to guide, motivate, and direct a team. include essential capabilities such as negotiation, resilience, communication, problem solving, critical thinking, and interpersonal skills.

**Dealing with people**
A project manager applies leadership skills and qualities when working with all project stakeholders, including the project team, the steering team, and project sponsors.

**Qualities and skills of a leader**
- Being a visionary (help to describe the products, goals, and objectives of the project);
- Being optimistic and positive;
- Being collaborative;
- Being respectful, courteous, friendly, kind, honest, trustworthy, loyal, and ethical;
- Exhibiting integrity and being culturally sensitive, courageous, a problem solver, and decisive;
- Giving credit to others where due;
- Being a life-long learner who is results- and action-oriented;

## The Project Manager Competency

 **LEADERSHIP** (Qualities and skills of a leader)
- **Focusing on the important things, including:**
    - Continuously prioritizing work by reviewing and adjusting as necessary;
    - Finding and using a prioritization method that works for them and the project;
    - Differentiating high-level strategic priorities
    - Maintaining vigilance on primary project constraints;
    - Remaining flexible on tactical priorities; and
    - Being able to sift through massive amounts of information to obtain the most important information.
    - Having a holistic and systemic view of the project, taking into account internal and external factors equally;
    - Being able to apply critical thinking
    - Being able to build effective teams, be service-oriented, and have fun and share humor effectively with team members.

### The Project Manager Competency

**LEADERSHIP** (Qualities and skills of a leader)

- **Managing relationships and conflict by:**
  - Building trust;
  - Satisfying concerns;
  - Seeking consensus;
  - Balancing competing and opposing goals;
  - Applying persuasion, negotiation, compromise, and conflict resolution skills;
  - Developing and nurturing personal and professional networks;
  - Taking a long-term view that relationships are just as important as the project; and
  - Continuously developing and applying political acumen.

- **Communicating by:**
  - Spending sufficient time communicating (90% of time)
  - Managing expectations;
  - Accepting feedback graciously;
  - Giving feedback constructively; and
  - Asking and listening.

## The Project Manager Competency

### LEADERSHIP
**Politics, Power, and Getting things done.**
Top project managers will work to acquire the power and authority they need within the boundaries of organizational policies, protocols, and procedures rather than wait for it to be granted.
- Positional (formal, authoritative, legitimate);
- Informational (e.g., control of gathering or distribution);
- Referent (e.g., respect or admiration others hold for the individual, credibility gained);
- Situational (e.g., gained due to unique situation such as a specific crisis);
- Personal or charismatic (e.g., charm, attraction);
- Relational (e.g., participates in networking, connections, and alliances);
- Expert (e.g., skill, information possessed; experience, training, education, certification);
- Reward-oriented (e.g., ability to give praise, monetary or other desired items);
- Punitive or coercive (e.g., ability to invoke discipline or negative consequences);
- Ingratiating (e.g., application of flattery or other common ground to win favor or cooperation);
- Pressure-based (e.g., limit freedom the purpose of gaining compliance to desired action);
- Guilt-based (e.g., imposition of obligation or sense of duty);
- Persuasive (e.g., ability to provide arguments that move people to a desired course of action); and
- Avoiding (e.g., refusing to participate).

## Manager vs Leader

**Project managers need to employ both leadership and management in order to be successful.**

| Management | Leadership |
|---|---|
| - Direct using positional power | - Guide, influence, and collaborate using relational power |
| - Maintain | - Develop |
| - Administrate | - Innovate |
| - Focus on systems and structure | - Focus on relationships with people |
| - Rely on control | - Inspire trust |
| - Focus on near-term goals | - Focus on long-range vision |
| - Ask how and when | - Ask what and why |
| - Focus on bottom line | - Focus on the horizon |
| - Accept status quo | - Challenge status quo |
| - Do things right | - Do the right things |
| - Focus on operational issues and problem solving | - Focus on vision, alignment, motivation, and inspiration |

## Leadership styles

**Laissez-faire:** allowing the team to make their own decisions and establish their own goals

**Transactional:** focus on goals, feedback, and accomplishment to determine rewards; management by exception);

**Servant leader:** demonstrates commitment to serve and put other people first; focuses on other people's growth, learning, development, autonomy, and well-being; concentrates on relationships, community and collaboration;

**Transformational:** empowering followers through idealized attributes and behaviors, inspirational motivation, encouragement for innovation and creativity, and individual consideration

**Charismatic:** able to inspire; is high-energy, enthusiastic, self-confident; holds strong convictions

**Interactional:** a combination of transactional, transformational, and charismatic

Project Manager

## Perform Integration

💡 **Integration is a critical skill for project managers.**

Integration and execution of the strategy.
**When working with the project sponsor to understand the strategic objectives and ensure the alignment of the project objectives and results with those of the portfolio, program, and business areas.**

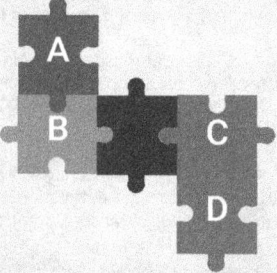

Integration of processes, knowledge, and people.
**By guiding the team to work together to focus on what is really essential at the project level.**

## Perform Integration

**Perform Integration at process level**
it is clear that a project has a small chance of meeting its objective when the project manager fails to integrate the project processes

**Integration at the cognitive level**
The PM should be proficient in all of the Project Management Knowledge Areas. Also applies experience, insight, leadership, and technical and business management skills to the project.

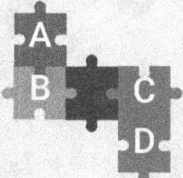

**Integration at the context level**
New technologies, Social networks, multicultural aspects, virtual teams, and new values are part of the new reality of projects.
The project manager considers the implications of this context in communications planning and knowledge management for guiding the project team.

**Integration and Complexity**
complexity are defined as: **System behavior.** ,**Human behavior.** Or **Ambiguity**
The project manager should examine the characteristics or properties that make the project complex an identify key areas when planning, managing, and controlling the project to ensure integration.

Have You Any
# question ?

| What | Why | When | Where | How | who |

# 4. PROJECT
## INTEGRATION MANAGEMENT

Presented by :
**Nasser Al Mohimeed**
PMO Director, ISO 21500 Lead Project Manager
Certified Project Managers Trainer

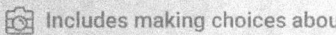 Project Integration Management

## Project Integration Management

Includes the processes and activities to identify, define, combine, unify, and coordinate the various processes and project management activities within the Project Management Process Groups.

**Includes making choices about:**

- Resource allocation.
- Balancing competing demands.
- Examining any alternative approaches.
- Tailoring the processes to meet objectives.
- Managing the interdependencies among the Project Management Knowledge Areas

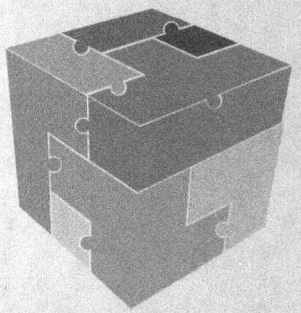

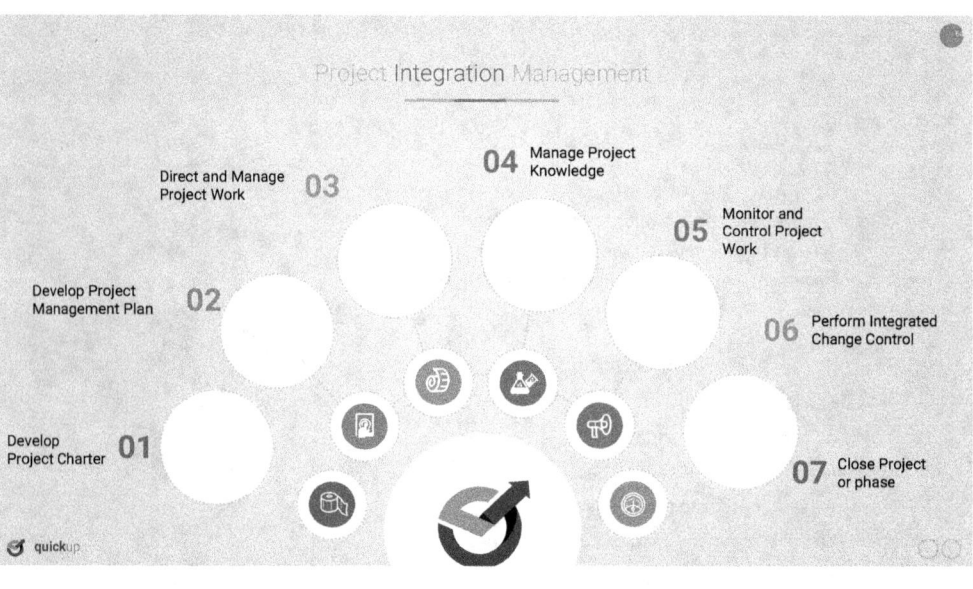

## Project Integration Managemet

| Knowledge Areas | Project Management Process Groups ||||| 
| | Initiating | Planning | Executing | Monitoring and Controlling | Closing |
|---|---|---|---|---|---|
| **Project Integration Management** | 4.1 Develop Project Charter | 4.2 Develop Project Management Plan | 4.3 Direct and Manage Project Work<br>4.4 Manage Project Knowledge | 4.5 Monitor &Control Project Work<br>4.6 Perform Integrated Change Control | 4.7 Close Project or phase |

## Key concepts for Project Integration Management

Project Management it is the specific responsibility of the project manager and it can' be delegated or transferred.

Project manager combines the results from all the other Knowledge Areas to provide an overall view of the project.

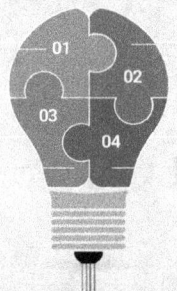

Projects and project management are integrative by nature

The project manager is ultimately responsible for the project as a whole

## Key concepts for Project Integration Management

**Project Integration Management is about:**
- Ensuring due dates of deliverables, life cycle, and benefits realization plan are aligned;
- Providing a project management plan to achieve the project objectives;
- Ensuring creation and use of appropriate knowledge to and from the project;
- Managing project performance and changes to the project activities;
- Making integrated decisions regarding key changes impacting the project;
- Measuring and monitoring progress and taking appropriate action;
- Collecting, analyzing and communicating project information to relevant stakeholders;
- Completing all the work of the project and formally closing each phase, contract, and the project as a whole
- Managing phase transitions when necessary.

## Key concepts for Project Integration Management

**Tailoring consideration**
Because each project is unique, the project manager may need to tailor the way that Project Integration Management

- Project life cycle
- Development life cycle.
- Management approaches
- Knowledge management.
- Change
- Governance
- Lessons learned
- Benefits

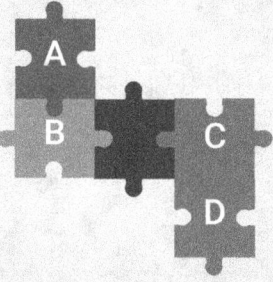

**Agile/ Adaptive Consideration**
- ✓ Iterative and agile approaches promote the engagement of team members as local domain experts in integration management.

- ✓ The team members determine how plans and components should integrate.

- ✓ control of the detailed product planning and delivery is delegated to the team.

- ✓ The project manager's focus is on building a collaborative decision-making environment

## 4.1 Develop Project Charter

**DEVELOP PROJECT CHARTER** the **process** of developing a document that formally authorizes the existence of a project and provides the project manager with the authority to apply organizational resources to project activities..

**THE KEY BENEFIT** it provides a direct link between the project and the strategic objectives of the organization, creates a formal record of the project, and shows the organizational commitment to the project.

## Develop Project Charter

### Develop Project Charter

**Inputs**
1. Business documents
   - Business case
   - Benefits management plan
2. Agreements
3. EEF
4. OPA

**Inputs Tools & Techniques Outputs**
1. Expert judgment
2. Data gathering
   - Brainstorming
   - Focus groups
   - Interviews
3. Interpersonal and team skills
   - Conflict management
   - Facilitation
   - Meeting management
4. Meetings

**Outputs**
.1 Project charter
.2 Assumption log

## 4.1 Develop Project Charter — Input

**01 Business documents**
- Business case
  necessary information from a business standpoint to determine whether or not the project is worth the required investment
  Created as a result from (Market demand, Organizational need, Customer request, Legal requirement, Social need)

**02 Agreements**
- They are used to define initial intentions for a project.
- Agreements May take the form of contracts, memorandums of understanding (MOUs), service level agreements (SLA), letters of agreement, letters of intent, verbal agreements, email, or other written agreements.
- A contract is used when a project is being performed for an external customer.

**03 Enterprise Environmental Factor**

**04 Organization Process Asset**

## 4.1 Develop Project Charter — Tools & Techniques - 1

**① Expert judgment**

- Defined as judgment provided based upon expertise in an application area, Knowledge Area, discipline, industry, etc., as appropriate for the activity being performed.

- Such expertise may be provided by any group or person with specialized education, knowledge, skill, experience, or training.

- For this process, expertise should be considered from individuals or groups with specialized knowledge of or training in the following topics:
    - Organizational strategy.
    - Benefits management.
    - Technical knowledge of the industry and focus area of the project.
    - Duration and budget estimation.
    - Risk identification.

## 4.1 Develop Project Charter    Tools & Techniques - 2

**Data gathering**

Brainstorming
- Is used to identify a list of ideas in a short period of time.
- It is conducted in a group environment and is led by a facilitator.
- Brainstorming comprises two parts: idea generation and analysis.
- Brainstorming can be used to gather data and solutions or ideas from stakeholders, subject matter experts, and team members when developing the project charter.

Focus group
- Bring together stakeholders and subject matter experts to learn about the perceived project risk, success criteria, and other topics in a more conversational way than a one-on-one interview.

Interviews
- are used to obtain information on high-level requirements, assumptions or constraints, approval criteria, and other information from stakeholders by talking directly to them.

## 4.1 Develop Project Charter **Tools & Techniques - 3**

**Interpersonal and team skills**

Conflict management
- Can be used to help bring stakeholders into alignment on the objectives, success criteria, high-level requirements, project description, summary milestones, and other elements of the charter.

Facilitation
- The ability to effectively guide a group event to a successful decision, solution, or conclusion.
- A facilitator ensures that there is effective participation, that participants achieve a mutual understanding, that all contributions are considered, that conclusions or results have full buy-in according to the decision process established for the project, and that the actions and agreements achieved are appropriately dealt with afterward.

Meeting management
- includes preparing the agenda, ensuring that a representative for each key stakeholder group is invited, and preparing and sending the follow-up minutes and actions.

**Meetings**
meetings are held with key stakeholders to identify the project objectives, success criteria, key deliverables, high-level requirements, summary milestones, and other summary information.

## 4.1 Develop Project Charter — Output

**01 Project Charter**
- The project charter is the document issued by the project initiator or sponsor that formally authorizes the existence of a project and provides the project manager with the authority to apply organizational resources to project activities

- Ensures a common understanding by the stakeholders of the key deliverables, milestones, and the roles and responsibilities of everyone involved in the project

**02 Assumption Log**
- High-level strategic and operational assumptions and constraints
- Lower-level activity and task assumptions (technical specifications, estimates, the schedule, risks)
- The assumption log is used to record all assumptions and constraints throughout the project life cycle.

## Project Charter

Project charter documents the high-level information on the project such as:

- Project purpose;
- Measurable project objectives and related success criteria;
- High-level requirements;
- High-level project description, boundaries, and key deliverables;
- Overall project risk;
- Summary milestone schedule;
- Preapproved financial resources;
- Key stakeholder list;
- Project approval requirements
- Project exit criteria Assigned project manager, responsibility, and authority level; and
- Name and authority of the sponsor or other person(s) authorizing the project charter.

## 4.2 Develop Project Management Plan

**DEVELOP PROJECT MANAGEMENT PLAN**
The process of defining, preparing, and coordinating all plan components and consolidating them into an integrated project management plan..

**THE KEY BENEFIT** is the production of a comprehensive document that defines the basis of all project work and how the work will be performed.

This process is performed once or at predefined points in the project.

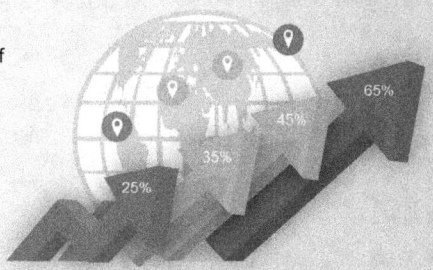

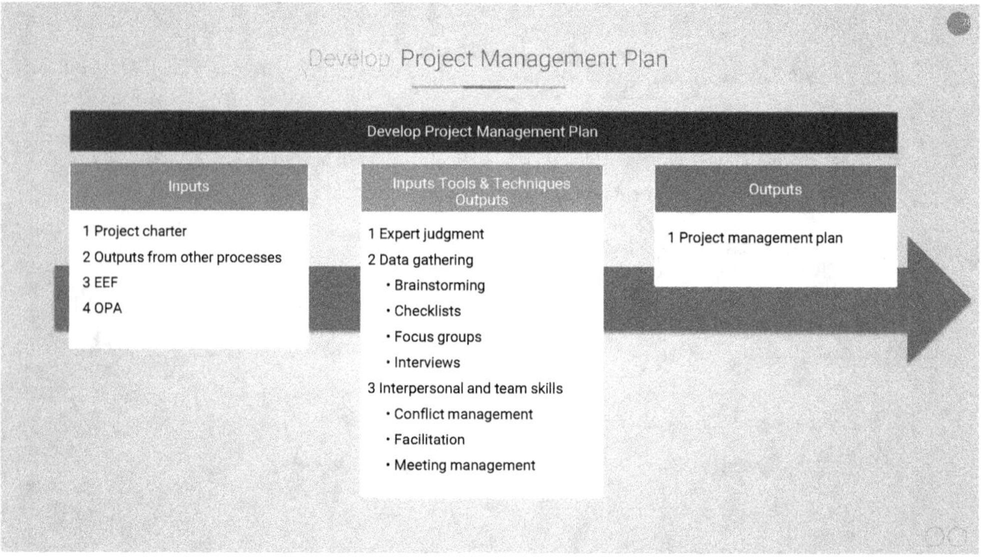

# Develop Project Management Plan

## Develop Project Management Plan

**Inputs**
1 Project charter
2 Outputs from other processes
3 EEF
4 OPA

**Tools & Techniques**
1 Expert judgment
2 Data gathering
 • Brainstorming
 • Checklists
 • Focus groups
 • Interviews
3 Interpersonal and team skills
 • Conflict management
 • Facilitation
 • Meeting management

**Outputs**
1 Project management plan

## 4.2 Develop Project Management Plan — Input

**01 Project charter**

**02 Outputs from other processes**
- Subsidiary plans
- All baselines

**03 Enterprise Environmental Factor**

**04 Organization Process Asset**

## 4.2 Develop Project Management Plan — Tools & Techniques

- **Expert judgment**

- **Data gathering**
  - Brainstorming
  - Focus group
  - Interviews
  - **Checklists**

- **Interpersonal and team skills**
  - Conflict management
  - Facilitation
  - Meeting management

- **Meetings**

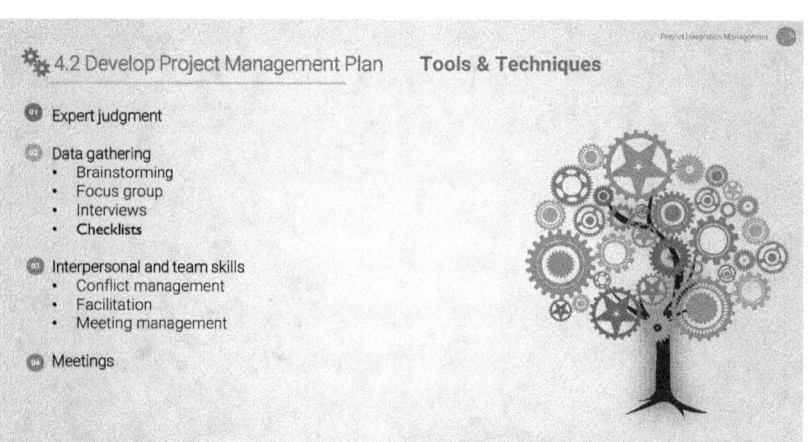

## 4.1 Develop Project Charter — Output

**Project Management Plan** is the document that describes how the project will be executed, monitored and controlled, and closed. It integrates and consolidates all of the subsidiary management plans and baselines, and other information necessary to manage the project

 Project baselines:
- Scope baseline.
- Schedule baseline.
- Cost baseline.

 Additional components as:
- Change management plan
- Configuration management plan
- Management reviews

 Subsidiary plans as:
- Scope management plan
- Requirements management plan
- Schedule management plan
- Cost management plan
- Quality management plan
- Process improvement plan

## Project Management Plan

1. Scope management plan
2. Requirements management plan
3. Schedule management plan
4. Cost management plan
5. Quality management plan
6. Resource management plan
7. Communications management plan
8. Risk management plan
9. Procurement management plan
10. Stakeholder engagement plan
11. Change management plan
12. Configuration management plan
13. Scope baseline
14. Schedule baseline
15. Cost baseline
16. Performance measurement baseline
17. Project life cycle description
18. Development approach

## Project Documents

1. Activity attributes
2. Activity list
3. Assumption log
4. Basis of estimates
5. Change log
6. Cost estimates
7. Cost forecasts
8. Duration estimates
9. Issue log
10. Lessons learned register
11. Milestone list
12. Physical resource assignments
13. Project calendars
14. Project communications
15. Project schedule
16. Project schedule network diagram
17. Project scope statement
18. Project team assignments
19. Quality control measurements
20. Quality metrics
21. Quality report
22. Requirements documentation
23. Requirements traceability matrix
24. Resource breakdown structure
25. Resource calendars
26. Resource requirements
27. Risk register
28. Risk report
29. Schedule data
30. Schedule forecasts
31. Stakeholder register
32. Team charter
33. Test and evaluation documents

## 4.3 Direct and Manage Project Work

**DIRECT AND MANAGE PROJECT WORK**
The process of leading and performing the work defined in the project management plan and implementing approved changes to achieve the project's objectives.

**THE KEY BENEFIT** is that it provides overall management of the project work and deliverables, thus improving the probability of project success

This process is performed throughout the project..

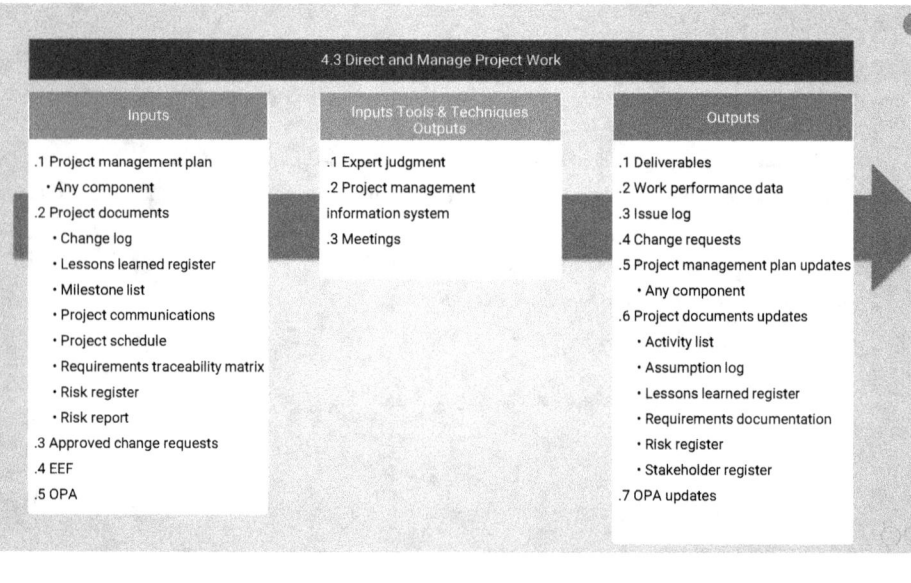

## 4.3 Direct and Manage Project Work — Input

1. **Project management plan**
2. **Project documents**
   - Change log
   - Lessons learned register
   - Milestone list
   - Project communications
   - Project schedule
   - Requirements traceability matrix
   - Risk register
   - Risk report
3. **Approved change requests** an output of the Perform Integrated Change Control process,
4. **EEFs.**
5. **OPA.**

## 4.3 Direct and Manage Project Work — Tools & Techniques

- **Expert judgment**

- **Project Management Information System(PMIS)**
  Is part of the environmental factors, provides access to tools
  - scheduling tool
  - configuration management system
  - information collection and distribution system
  - Interfaces to other online automated systems.

- **Meetings**

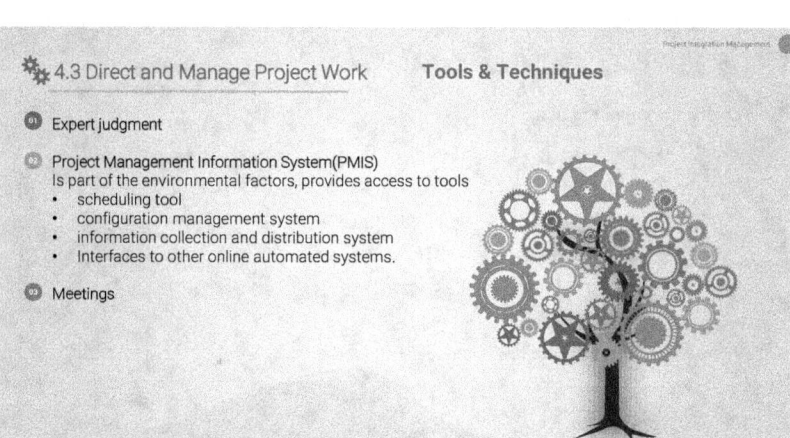

## 4.3 Direct and Manage Project Work — Output - 1

**01 Deliverable:** deliverable is any unique and verifiable product, result, or capability to perform a service that is required to be produced to complete a process, phase, or project.

**02 Work performance data:** are the raw observations and measurements identified during activities being performed to carry out the project work.

**03 Issue log:** is a project document where all the issues are recorded and tracked Project communications

## 4.3 Direct and Manage Project Work — Output - 2

- **Change Requests**
  - **Corrective action.** An intentional activity that realigns the performance of the project work with the project management plan.
  - **Preventive action.** An intentional activity that ensures the future performance of the project work is aligned with the project management plan.
  - **Defect repair.** An intentional activity to modify a nonconforming product or product component.
  - **Updates.** Changes to formally controlled project documents, plans, etc., to reflect modified or additional ideas or content.
- **Project management plan updates**
- **Project document updates**
- **Organizational process assets updates**

## 4.4 Manage Project Knowledge

**MANAGE PROJECT KNOWLEDGE** the process of using existing knowledge and creating new knowledge to achieve the project's objectives and contribute to organizational learning

**THE KEY BENEFIT** are that prior organizational knowledge is leveraged to produce or improve the project outcomes, and knowledge created by the project is available to support organizational operations and future projects or phases.

## 4.4 Manage Project Knowledge

### Inputs

1 Project management plan
 • All components
2 Project documents
 • Lessons learned register
 • Project team assignments
 • Resource breakdown structure
 • Source selection criteria
 • Stakeholder register
3 Deliverables
4 EEF
5 OPA

### Tools & Techniques

1 Expert judgment
2 Knowledge management
3 Information management
4 Interpersonal and team skills
 • Active listening
 • Facilitation
 • Leadership
 • Networking
 • Political awareness

### Outputs

1 Lessons learned register
2 Project management plan updates
 • Any component
3 OPA update

## 4.4 Manage Project Knowledge — Input

1. **Project management plan.**
2. **Project documents.**
   - Lessons learned register
   - Project team assignments
   - Resource breakdown structure
   - Source selection criteria
   - Stakeholder register
3. **Deliverables.**
   - deliverable is any unique and verifiable product, result, or capability to perform a service that is required to be produced to complete a process, phase, or project.
   - Deliverables are typically the outcomes of the project and can include components of the project management plan.
4. **EEFs.**
5. **OPA.**

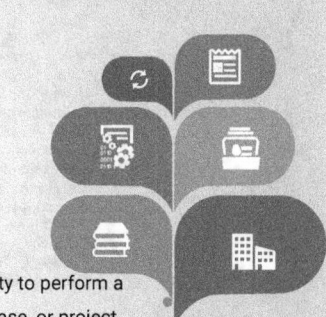

## 4.4 Manage Project Knowledge — Tools & Techniques - 1

**Expert judgment**
Knowledge management tools and techniques connect people so they can work together to create new knowledge, share tacit knowledge, and integrate the knowledge of diverse team members.
- **Networking**, including informal social interaction and online social networking.
- **Communities of practice** and special interest groups.
- **Meetings**, including virtual meetings where participants can interact using communications technology.
- **Work shadowing and reverse shadowing**.
- **Discussion forums** such as focus groups;
- **Knowledge-sharing events** such as seminars and conferences;
- **Workshops**, including problem-solving sessions and learning reviews designed to identify lessons learne
- **Storytelling**;
- Creativity and ideas management techniques.
- Knowledge fairs and cafés.
- Training that involves interaction between learners.

- Information management tools and techniques
  ➢ They are used to create and connect people to information.
  ➢ They are effective for sharing simple, unambiguous, codified explicit knowledge.

## 4.4 Manage Project Knowledge — Tools & Techniques - 2

**Interpersonal and team skills**
- Active listening.
  Helps reduce misunderstandings and improves communication and knowledge sharing.
- Facilitation.
  Helps effectively guide a group to a successful decision, solution, or conclusion.
- Leadership.
  It is used to communicate the vision and inspire the project team to focus on the appropriate knowledge and knowledge objectives.
- Networking.
  Allows informal connections and relations among project stakeholders to be established and creates the conditions to share tacit and explicit knowledge.
- Political awareness.
  Helps the project manager to plan communications based on the project environment as well as the organization's political environment.

## 4.4 Manage Project Knowledge — Output

**01  Lessons learned register**
can include
- the category and description of the situation
- the impact, recommendations, and proposed actions associated with the situation.
- record challenges, problems, realized risks and opportunities, or other content as appropriate.

**02  Project management plan updates**
**03  OPA updates**

## 4.5 Monitor and Control Project Work

 **MONITOR AND CONTROL PROJECT WORK** the process of tracking, reviewing, and reporting the overall progress to meet the performance objectives defined in the project management plan

**THE KEY BENEFIT** are that it allows stakeholders to understand the current state of the project, to recognize the actions taken to address any performance issues, and to have visibility into the future project status with cost and schedule forecasts.

## 4.5 Monitor and Control Project Work

**Inputs**
1. Project management plan
2. Project documents
   - Assumption log
   - Basis of estimates
   - Cost forecasts
   - Issue log
   - Lessons learned register
   - Milestone list
   - Quality reports
   - Risk register
   - Risk report
   - Schedule forecasts
3. Work performance information
4. Agreements
5. Enterprise environmental factors
6. Organizational process assets

**Tools & Techniques**
1. Expert judgment
2. Data analysis
   - Alternatives analysis
   - Cost-benefit analysis
   - Earned value analysis
   - Root cause analysis
   - Trend analysis
   - Variance analysis
3. Decision making
4. Meetings

**Outputs**
1. Work performance reports
2. Change requests
3. Project management plan updates
4. Project documents updates
   - Cost forecasts
   - Issue log
   - Lessons learned register
   - Risk register
   - Schedule forecasts

# 4.5 Monitor and Control Project Work — Input - 1

- **Project management plan**
- **Project documents**
  - Assumption log
  - Basis of estimates
  - Cost forecasts
  - Issue log
  - Lessons learned register
  - Milestone list
  - Quality reports
  - Risk register
  - Risk report
  - Schedule forecasts

## 4.5 Monitor and Control Project Work    Input - 2

**Work performance information**
- It is gathered through work execution and passed to the controlling processes.
- To become work performance information, the work performance data are compared with the project management plan components, project documents, and other project variables.
- This comparison indicates how the project is performing.
- Specific work performance metrics for scope, schedule, budget, and quality are defined at the start of the project as part of the project management plan.
- Performance data are collected during the project through the controlling processes and compared to the plan and other variables to provide a context for work performance.

**Agreements**

**EEFs**

OPA

## 4.5 Monitor and Control Project Work — Tools & Techniques - 1

- Expert judgment
- Data analysis
    - **Alternatives analysis.** is used to select the corrective actions or a combination of corrective and preventive actions to implement when a deviation occurs.
    - **Cost-benefit analysis** helps to determine the best corrective action in terms of cost in case of project deviations.
    - **Earned value analysis.** provides an integrated perspective on scope, schedule, and cost performance.
    - **Root cause analysis** focuses on identifying the main reasons of a problem. It can be used to identify the reasons for a deviation and the areas the project manager should focus on in order to achieve the objectives of the project.

## 4.5 Monitor and Control Project Work — Tools & Techniques - 2

- Data analysis
  - Trend analysis.
  - It is used to forecast future performance based on past results.
  - It looks ahead in the project for expected slippages and warns the project manager ahead of time that there may be problems later in the schedule if established trends persist.
  - This information is made available early enough in the project timeline to give the project team time to analyze and correct any anomalies.
  - The results of trend analysis can be used to recommend preventive actions if necessary.

## 4.5 Monitor and Control Project Work — Tools & Techniques - 3

### Data analysis

Variance analysis:

- Reviews the differences (or variance) between planned and actual performance. This can include duration estimates, cost estimates, resources utilization, resources rates, technical performance, and other metrics.
- It may be conducted in each Knowledge Area based on its particular variables.
- In Monitor and Control Project Work, the variance analysis reviews the variances from an integrated perspective considering cost, time, technical, and resource variances in relation to each other to get an overall view of variance on the project.
- This allows for the appropriate preventive or corrective actions to be **initiated**.

### Decision making

It can include making decisions based on unanimity, majority, or plurality.

### Meetings

## 4.5 Monitor and Control Project Work — Output

- **01** Work performance reports
- **02** Change requests
- **03** Project management plan updates
- **04** Project documents updates
  - Cost forecasts
  - Issue log
  - Lessons learned register
  - Risk register
  - Schedule forecasts

## 4.6 Perform Integrated Change Control

the process of **reviewing all change requests; approving changes and managing changes to deliverables, project documents, and the project management plan; and communicating the decisions.**

This process reviews all requests for changes to project documents, deliverables, or the project management plan and determines the resolution of the change requests.

THE KEY BENEFIT **that it allows for documented changes within the project to be considered in an integrated manner while addressing overall project risk, which often arises from changes made without consideration of the overall project objectives or plans.**

## 4.6 Perform Integrated Change Control

### Inputs

.1 Project management plan
   - Change management plan
   - Configuration management plan
   - Scope baseline
   - Schedule baseline
   - Cost baseline

.2 Project documents
   - Basis of estimates
   - Requirements traceability matrix
   - Risk report

.3 Work performance reports

.4 Change requests

.5 Enterprise environmental factors

.6 Organizational process assets

### Tools & Techniques

.1 Expert judgment

.2 Change control tools

.3 Data analysis
   - Alternatives analysis
   - Cost-benefit analysis

.4 Decision making
   - Voting
   - Autocratic decision making
   - Multi-criteria decision analysis

.5 Meetings

### Outputs

.1 Approved change requests

.2 Project management plan updates
   - Any component

.3 Project documents updates
   - Change log

## 4.6 Perform Integrated Change Control — Input - 1

**Project management plan**
- Change management plan: provides the direction for managing the change control process
- Configuration management plan: describes the configurable items of the project and identifies the items that will be recorded and updated
- Scope baseline: which contains the procedures for scope changes.
- Schedule baseline : which is used to assess the impact of the changes in the project schedule.
- Cost baseline : which is used to assess the impact of the changes to the project cost.

## 4.6 Perform Integrated Change Control — Input - 2

**Project documents.**
- Basis of estimates
- Requirements traceability matrix
- Risk report

**Work performance reports.**
Reports of particular interest to the Perform Integrated Change Control process include resource availability, schedule and cost data, earned value reports, and burn-up or burn-down charts.

**Change requests.**

**EEFs.**

**OPA.**

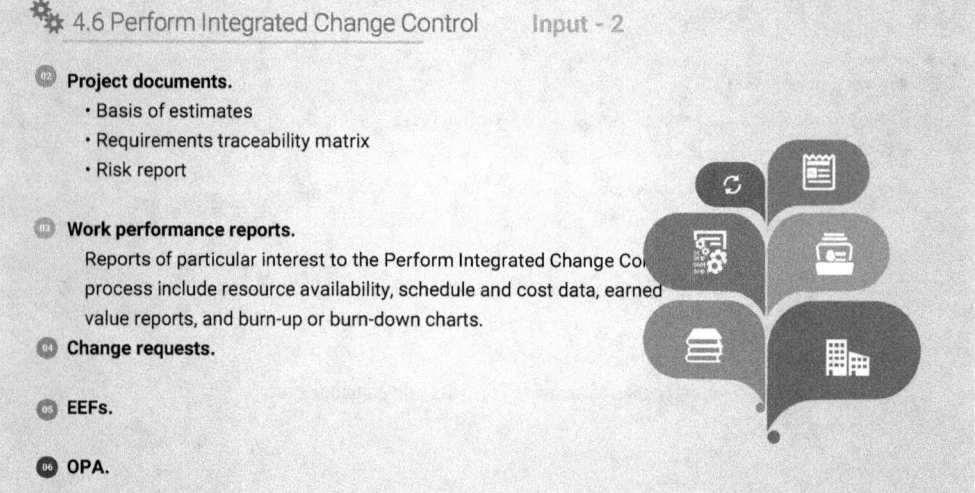

## 4.6 Perform Integrated Change Control — Tools & Techniques 1

**EXPERT JUDGMENT**
In addition to the project management team's expert judgment, stakeholders may be asked to provide their expertise and maybe asked to sit on the change control board (CCB).

**CHANGE CONTROL TOOLS**
- Tools are used to manage the change requests and the resulting decisions.

- Tools should support the following configuration management activities:
    - Identify configuration item.
    - Record and report configuration item status.
    - Perform configuration item verification and audit.

- Tools should support the following change management activities as well:
    - Identify changes.
    - Document changes.
    - Decide on changes.
    - Track changes.

## 4.6 Perform Integrated Change Control — Tools & Techniques - 2

**DATA ANALYSIS**
- Alternatives analysis.
- Cost-benefit analysis.

**DECISION MAKING**
- Voting.
  - ➤ Voting can take the form of unanimity, majority, or plurality to decide on whether to accept, defer, or reject change requests.
- Autocratic decision making.
  - ➤ one individual takes the responsibility for making the decision for the entire group.
- Multi-criteria decision analysis.
  - ➤ This technique uses a decision matrix to provide a systematic analytical approach to evaluate the requested changes according to a set of predefined criteria.

**MEETINGS**

## 4.6 Perform Integrated Change Control — Output

**① APPROVED CHANGE REQUESTS**
- Approved change requests will be implemented through the Direct and Manage Project Work process.
- Deferred or rejected change requests are communicated to the person or group requesting the change.
- The disposition of all change requests are recorded in the change log as a project document update.

**② PROJECT MANAGEMENT PLAN UPDATES**

**③ PROJECT DOCUMENTS UPDATES.**

## 4.7 Close Project or Phase

 the process of finalizing all activities for the project, phase, or contract.

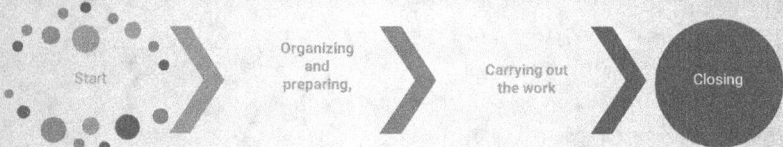

THE KEY BENEFIT are the project or phase information is archived, the planned work is completed, and organizational team resources are released to pursue new endeavors. This process is performed once or at predefined points in the project.

## 4.7 Close Project or Phase

 the project manager reviews the project management plan to ensure that all project work is completed and that the project has met its objectives. The activities necessary for the administrative closure of the project or phase

 Actions and activities necessary to satisfy completion or exit criteria for the phase or project
- Making certain that all documents and deliverables are up-to-date and that all issues are resolved;
- Confirming the delivery and formal acceptance of deliverables by the customer;
- Ensuring that all costs are charged to the project;
- Closing project accounts;
- Reassigning personnel;
- Dealing with excess project material;
- Reallocating project facilities, equipment, and other resources; and
- Elaborating the final project reports as required by organizational policies

## 4.7 Close Project or Phase

Activities related to the completion of the contractual agreements applicable to the project or project phase:

- Confirming the formal acceptance of the seller's work,
- Finalizing open claims,
- Updating records to reflect final results, and
- Archiving such information for future use.
- Activities needed to:
- Collect project or phase records,
- Audit project success or failure,
- Manage knowledge sharing and transfer,
- Identify lessons learned, and
- Archive project information for future use by the organization.

Actions and activities necessary to transfer the project's products, services, or results to the next phase or to production and/or operations:
- Collecting any suggestions for improving or updating the policies and procedures of the organization, and sending them to the appropriate organizational unit.
- Measuring stakeholder satisfaction.

## 4.7 Close Project or Phase

**Inputs**
1. Project charter
2. Project management plan
3. Project documents
   - Assumption log
   - Basis of estimates
   - Change log
   - Issue log
   - Lessons learned register
   - Milestone list
   - Project communications
   - Quality control measurements
   - Quality reports
   - Requirements documentation
   - Risk register
   - Risk report
4. Accepted deliverables
5. Business documents
   - Business case
   - Benefits management plan
6. Agreements
7. Procurement documentation
8. Organizational process assets

**Tools & Techniques**
.1 Expert judgment
.2 Data analysis
   - Document analysis
   - Regression analysis
   - Trend analysis
   - Variance analysis
.3 Meetings

**Outputs**
.1 Project documents updates
   - Lessons learned register
.2 Final product, service, or result transition
.3 Final report
.4 Organizational process assets updates

## 4.7 Close Project or Phase — Input - 1

**① PROJECT CHARTER**

**② PROJECT MANAGEMENT PLAN**

**③ PROJECT DOCUMENTS**
- Assumption log.
- Basis of estimates.
- Change log.
- Issue log.
- Lessons learned register.
- Milestone list.
- Project communications.
- Quality control measurements.
- Quality reports.
- Requirements documentation.
- Risk register.
- Risk reports.

## 4.7 Close Project or Phase — Input - 2

- **ACCEPTED DELIVERABLES.**

- **BUSINESS DOCUMENTS.**
  - Business case: documents the business need and the cost benefit analysis that justify the project.
  - Benefits management plan.
  - Outlines the target benefits of the project.
  - It is used to determine if the expected outcomes from the economic feasibility study used to justify the project occurred.
  - It is used to measure whether the benefits of the project were achieved as planned.

- **AGREEMENTS.**

- **PROCUREMENT DOCUMENTATION.**

- **OPA.**

## 4.7 Close Project or Phase — Tools & Techniques

**EXPERT JUDGMENT**

**DATA ANALYSIS**
- Document analysis.
  - Assessing available documentation will allow identifying lessons learned and knowledge sharing for future projects and organizational assets improvement.
- Regression analysis.
  - Analyzes the interrelationships between different project variables that contributed to the project outcomes to improve performance on future projects.
- Trend analysis.
  - It is used to validate the models used in the organization and to implement adjustments for future projects.
- Variance analysis.
  - It may be used to improve the metrics of the organization by comparing what was initially planned and the end result.

**MEETINGS**
Meetings may be face-to-face, virtual, formal, or informal. Types of meetings include but are not limited to close-out reporting meetings, customer wrap-up meetings, lessons learned meetings, and celebration meetings.

## 4.7 Close Project or Phase — Output

**01 PROJECT DOCUMENTS UPDATES**

**02 FINAL PRODUCT, SERVICE, OR RESULT TRANSITION**

**03 FINAL REPORT**
The final report provides a summary of the project performance. It can include information such as:
- Summary level description of the project or phase.
- Scope objectives, the criteria used to evaluate the scope, and evidence that the completion criteria were met.
- Quality objectives, the criteria used to evaluate the project and product quality, the verification and actual milestone delivery dates, and reasons for variances.
- Cost objectives, including the acceptable cost range, actual costs, and reasons for any variances.
- Summary of the validation information for the final product, service, or result.

**04 OPA UPDATES**

Have You Any
# question ?

| What | Why | When | Where | How | who |

# 5. PROJECT
## SCOPE MANAGEMENT

Presented by :
**Nasser Al Mohimeed**
PMO Director, ISO 21500 Lead Project Manager
Certified Project Managers Trainer

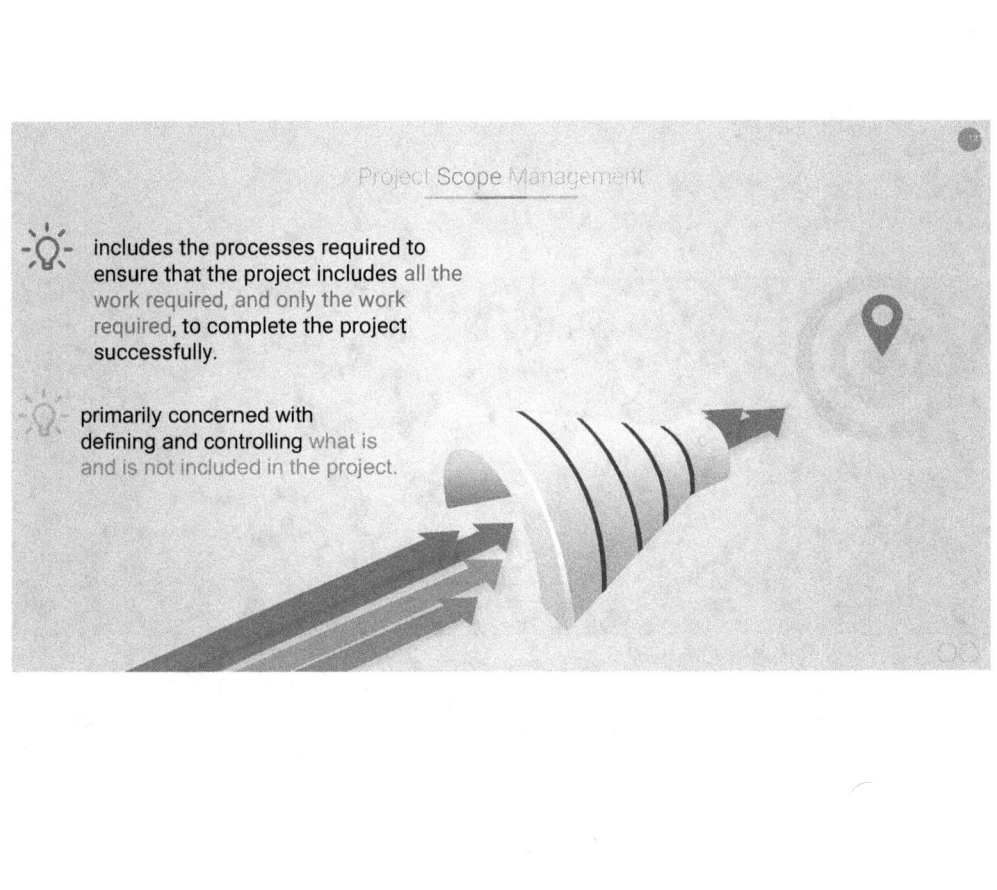

## Project Scope Management

- 💡 **includes the processes required to ensure that the project includes** all the work required, and only the work required, **to complete the project successfully.**

- 💡 **primarily concerned with defining and controlling** what is and is not included in the project.

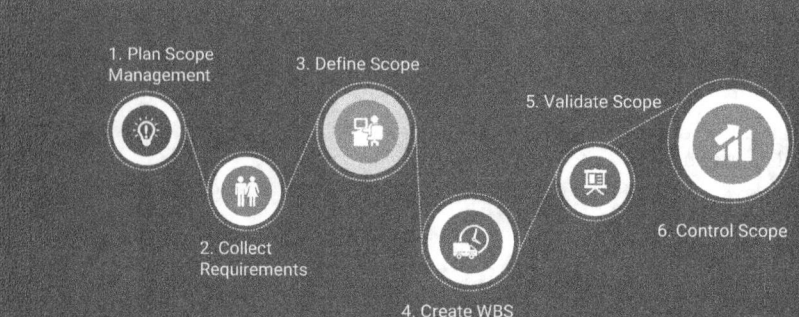

# Project Scope Management

| Initiating | Planning | Executing | Monitoring & Controlling | Closing |
|---|---|---|---|---|
| | 5.1 Plan Scope Management | | 5.5 Validate Scope | |
| | 5.2 Collect Requirements | | 5.6 Control Scope | |
| | 5.3 Define Scope | | | |
| | 5.4 Create WBS | | | |

## Key concepts for Project Scope Management

The term "scope" can refer to:

Product scope. The features and functions that characterize a product, service, or result.

Project scope. The work performed to deliver a product, service, or result with the specified features and functions. The term "project scope" is sometimes viewed as including product scope.

- ❖ Project scope completion is measured against the project management plan,
- ❖ product scope completion is measured against the product requirements

Project life cycles can be :

- ❖ Predictive life cycle
the project deliverables are defined at the beginning of the project and any changes to the scope are progressively managed.

- ❖ Adaptive or agile life cycle
the deliverables are developed over multiple iterations where a detailed scope is defined and approved for each iteration when it begins.

## Key concepts for Project Scope Management

**Projects with adaptive life cycles- agile**

- Respond to high levels of change
- Require ongoing stakeholder engagement.
- The overall scope of an adaptive project will be decomposed into a set of requirements and work to product backlog.
- At the beginning of an iteration, the team will work to determine how many of the highest-priority items on the backlog list can be delivered within the next iteration.
- Collect Requirements, Define Scope, and Create WBS; are repeated for each iteration.
- The sponsor and customer should be continuously engaged with the project to provide feedback on deliverables as they are created and to ensure that the product backlog reflects their current needs.

## Key concepts for Project Scope Management

**Projects with adaptive life cycles- agile**

- Validate Scope and Control Scope; are repeated for each iteration.
- The scope baseline for the project is the approved version of the project scope statement (WBS and WBS dictionary)
- A baseline can be changed only through formal change control procedures and is used as a basis for comparison while performing Validate Scope and Control Scope processes
- Use backlogs (including product requirements and user stories) to reflect their current needs.

## Key concepts for Project Scope Management

**TRENDS AND EMERGING PRACTICES IN PROJECT SCOPE MANAGEMENT**

- using business analysis **is important to organization to their competitive advantage by defining, managing, and controlling requirements activities.**

- Activities of business analysis **may start before a project is initiated and a project manager is assigned.**

- Requirements Management **(Eliciting, documenting, and managing stakeholder requirements) takes place within the Project Scope Management processes.**

- **Project Scope Management include** collaborating **with business analysis professionals to:**
    - Determine problems and identify business needs;
    - Identify and recommend viable solutions for meeting those needs;
    - Elicit, document, and manage stakeholder requirements in order to meet business and project objectives;
    - Facilitate the successful implementation of the product, service, or end result of the program or project.

## Key concepts for Project Scope Management

**TRENDS AND EMERGING PRACTICES IN PROJECT SCOPE MANAGEMENT**

 The business analyst is responsible on Requirement-related activities

 The project manager is responsible **for**
- ensuring that requirements-related work is accounted for in the project management plan
- Requirements-related activities are performed on time and within budget and deliver value.

 The relationship between a project manager and a business analyst should be a collaborative partnership.

 To successfully achieve project objectives, project managers and business analysts should understand each other's roles and responsibilities

## Key concepts for Project Scope Management

**TAILORING CONSIDERATIONS**

- Knowledge and requirements management. Does the organization have formal or informal knowledge and requirements management systems? What guidelines should the project manager establish for requirements to be reused in the future?

- Validation and control. Does the organization have existing formal or informal validation and control-related policies, procedures, and guidelines?

- Development approach. Does the organization use agile approaches in managing projects? Is the development approach iterative or incremental? Is a predictive approach used? Will a hybrid approach be productive?

- Stability of requirements. Are there areas of the project with unstable requirements? Do unstable requirements necessitate the use of lean, agile, or other adaptive techniques until they are stable and well defined?

- Governance. Does the organization have formal or informal audit and governance policies, procedures, and guidelines?

## Key concepts for Project Scope Management

### CONSIDERATIONS FOR AGILE/ADAPTIVE ENVIRONMENTS

In project with emerging requirements there is often a gap between the real business requirements and the business requirements that were originally stated.

Agile methods spend less time to define and agree on scope in the early stage
spend more time establishing the process for its ongoing discovery and refinement.

Agile methods purposefully build and review prototypes and release versions in order to refine the requirements. scope is defined and redefined throughout the project

## 5.1 Plan Scope Management

 **PLAN SCOPE MANAGEMENT** the process of creating a scope management plan that documents how the project and product scope will be defined, validated, and controlled.

**THE KEY BENEFIT** is that it provides guidance and direction on how scope will be managed throughout the project.

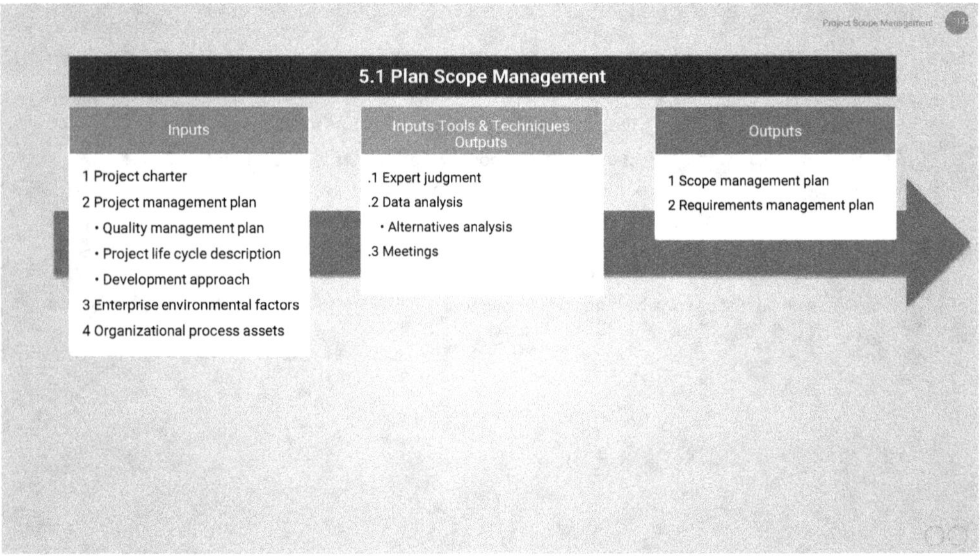

Project Scope Management
5.1 Plan Scope Management
Inputs
1 Project charter
2 Project management plan
• Quality management plan
• Project life cycle description
• Development approach
3 Enterprise environmental factors
4 Organizational process assets
Inputs Tools & Techniques Outputs
.1 Expert judgment
.2 Data analysis
• Alternatives analysis
.3 Meetings
Outputs
1 Scope management plan
2 Requirements management plan

## 5.1 Plan Scope Management — Input

**01 PROJECT CHARTER**

**02 PROJECT MANAGEMENT PLAN**

- **Quality management plan** organization's quality policy, methodologies, and standards are implemented on the project.
- **Project life cycle description.** the series of phases that a project passes
- **Development approach.** whether waterfall, iterative, adaptive, agile, or a hybrid development approach will be used.

**03 ENTERPRISE ENVIRONMENTAL FACTORS**

**04 ORGANIZATIONAL PROCESS ASSETS**

## 5.1 Plan Scope Management — Tools & Techniques

**01** Expert judgment

**02** Data Analysis
- **Alternatives analysis**

**03** Meetings

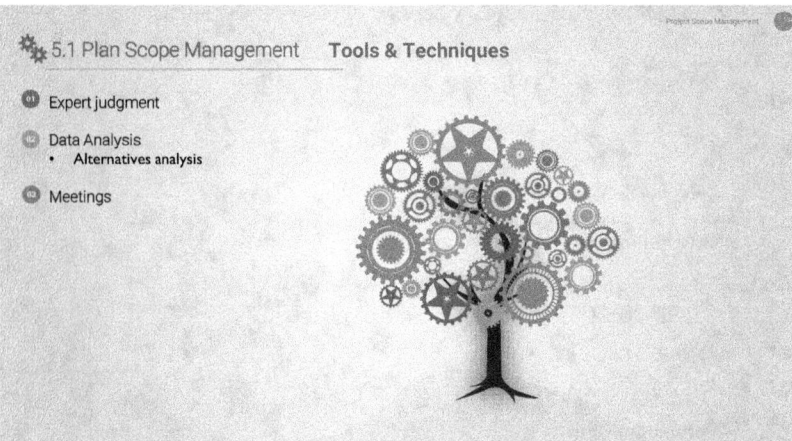

## 5.1 Plan Scope Management — Output 1

 **SCOPE MANAGEMENT PLAN**
Component of the project management plan that describes how the scope will be defined, developed, monitored, controlled, and validated; includes
- Process for preparing a project scope statement;
- Process that enables the creation of the WBS from the detailed project scope statement;
- Process that establishes how the scope baseline will be approved and maintained; and
- Process that specifies how formal acceptance of the completed project deliverables will be obtained.

## 5.1 Plan Scope Management — Output 2

 **REQUIREMENTS MANAGEMENT PLAN**
- ➤ Component of the project management plan that describes how project and product requirements will be analyzed, documented, and managed.
- ➤ some organizations refer to it as a business analysis plan.

- How requirements activities will be planned, tracked, and reported;
- Configuration management activities such as: how changes will be initiated; how impacts will be analyzed; how they will be traced, tracked, and reported; as well as the authorization levels required to approve these changes;
- Requirements prioritization process;
- Metrics that will be used and the rationale for using them; and
- Traceability structure that reflects the requirement attributes captured on the traceability matrix.

## 5.2 Collect Requirements

 COLLECT REQUIREMENTS is the process of determining, documenting, and managing stakeholder needs and requirements to meet objectives

THE KEY BENEFIT it provides the basis for defining the product scope and project scope. This process is performed once or at predefined points in the project
- Business requirements
- Stakeholder requirements
- Solution requirements
- Project requirements
- Quality requirements

## 5.2 Collect Requirements

**Inputs**
- .1 Project charter
- .2 Project management plan
  - Scope management plan
  - Requirements management plan
  - Stakeholder engagement plan
- .3 Project documents
  - Assumption log
  - Lessons learned register
  - Stakeholder register
- .4 Business documents
  - Business case
- .5 Agreements
- .6 Enterprise environmental factors
- .7 Organizational process assets

**Inputs Tools & Techniques Outputs**
- .1 Expert judgment
- .2 Data gathering
  - Brainstorming
  - Interviews
  - Focus groups
  - Questionnaires and surveys
  - Benchmarking
- .3 Data analysis
  - Document analysis
- .4 Decision making
  - Voting
  - Multicriteria decision analysis
- .5 Data representation
  - Affinity diagrams
  - Mind mapping
- .6 Interpersonal and team skills
  - Nominal group technique
  - Observation/conversation
  - Facilitation
- .7 Context diagram
- .8 Prototypes

**Outputs**
- .1 Requirements documentation
- .2 Requirements traceability matrix

## 5.2 Collect Requirements — Input 1

**PROJECT CHARTER**

**PROJECT MANAGEMENT PLAN**
- Scope management plan. **contains information on how the project scope will be defined and developed.**
- Requirements management plan. **how project requirements will be collected, analyzed, and documented.**
- Stakeholder engagement plan. **understand stakeholder communication requirements and the level of stakeholder engagement in order to assess and adapt to the level of stakeholder participation in requirements activities.**

**PROJECT DOCUMENTS**
- Assumption Log.
- Lessons learned register.
- Stakeholder Register.

## 5.2 Collect Requirements — Input 2

**BUSINESS DOCUMENTS**
A business document that can influence the Collect Requirements process is the business case, which can describe required, desired, and optional criteria for meeting the business needs.

**AGREEMENTS**

**ENTERPRISE ENVIRONMENTAL FACTORS**

**ORGANIZATIONAL PROCESS ASSETS**

## 5.2 Collect Requirements — Tools & Techniques 1

**DATA GATHERING**

Brainstorming. generate and collect multiple ideas related to project and product requirements.

Focus groups. bring together stakeholders and subject matter experts to learn about their expectations and attitudes about a proposed product, service, or result.

Interviews. Formal/ informal approach to elicit information from stakeholders by talking to them directly.

Questionnaires and surveys. written sets of questions designed to quickly accumulate information from a large number of respondents

Benchmarking comparing actual or planned products, to those of comparable organizations to identify best practices, generate ideas for improvement, and provide a basis for measuring performance

## 5.2 Collect Requirements — Tools & Techniques 2

**EXPERT JUDGMENT**

**DATA ANALYSIS**
Document analysis: reviewing and assessing any relevant documented information to elicit requirements by analyzing existing documentation and identifying information relevant to the requirements.

**DECISION MAKING**
Voting.
- Unanimity - everyone agrees on a single course of action.
- Majority – More than 50% of the members agree.
- Plurality. – Largest block in a group decides.

Autocratic decision making **one individual takes responsibility for making the decision**
Multi-criteria decision analysis **uses a decision matrix to provide a systematic analytical approach for establishing criteria to evaluate and rank many ideas.**

## 5.2 Collect Requirements — Tools & Techniques 2

**DATA REPRESENTATION**
- Affinity diagrams - large numbers of ideas to be classified into groups for review and analysis.
- Mind mapping - consolidates ideas created through individual brainstorming sessions into a single map to reflect commonality and differences in understanding and to generate new ideas.

**INTERPERSONAL AND TEAM SKILLS**
- Nominal group technique - brainstorming with a voting process used to rank the most useful ideas for further brainstorming or for prioritization.
- Observation/conversation - a direct way of viewing individuals in their environment and how they perform their jobs or tasks and carry out processes.
- Facilitation. Facilitation is used with focused sessions that bring key stakeholders
- together to define product requirements.

**Prototypes** providing a model of the expected product before actually building it, to obtaining early feedback on requirements

## 5.2 Collect Requirements — Tools & Techniques 2

**CONTEXT DIAGRAM**
visually depict the product scope by showing a business system (process, equipment, computer system, etc.), and how people and other systems (actors) interact with it

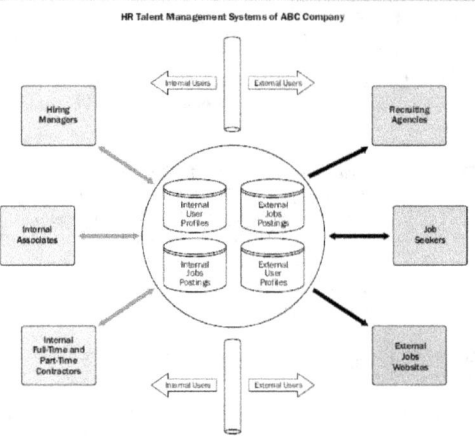

## 5.2 Collect Requirements — Output 1

### REQUIREMENTS DOCUMENTATION

Describes how individual requirements meet the business need for the project. Requirements may start out at a high level and become progressively more detailed as more information about the requirements is known.

- **Business requirements.** describe the higher-level needs of the organization as a whole, such as the business issues or opportunities, and reasons why a project has been undertaken.

- **Stakeholder requirements.** describe needs of a stakeholder or stakeholder group.

- **Solution requirements.** describe features, functions, and characteristics of the product, service, or result that will meet the business and stakeholder requirements.
  Functional Ex: actions, processes, data, and interactions that the product should execute
  Nonfunctional Ex: reliability, security, performance, safety, level of service, supportability, retention.

- **Transition and readiness requirements.** data conversion and training requirements, needed to transition from the current as-is state to the desired future state.

- **Project requirements.** Ex: include milestone dates, contractual obligations, constraints, etc.

- **Quality requirements.** Ex: include tests, certifications, validations, etc.

## 5.2 Collect Requirements — Output 2

**REQUIREMENTS TRACEABILITY MATRIX**
- Grid that links product requirements from their origin to the deliverables
- Ensure that each requirement adds business value by linking it to the business and project objectives.

### Requirements Traceability Matrix

| Project Name: | |
|---|---|
| Cost Center: | |
| Project Description: | |

| ID | Associate ID | Requirements Description | Business Needs, Opportunities, Goals, Objectives | Project Objectives | WBS Deliverables | Product Design | Product Development | Test Cases |
|---|---|---|---|---|---|---|---|---|
| 001 | 1.0 | | | | | | | |
|  | 1.1 | | | | | | | |
|  | 1.2 | | | | | | | |
|  | 1.2.1 | | | | | | | |
| 002 | 2.0 | | | | | | | |
|  | 2.1 | | | | | | | |

## 5.3 Define Scope

**DEFINE SCOPE** the process of developing a detailed description of the project and product.

**THE KEY BENEFIT** it describes the product, service, or result boundaries and acceptance criteria.
Describes how individual requirements meet the business need for the project.

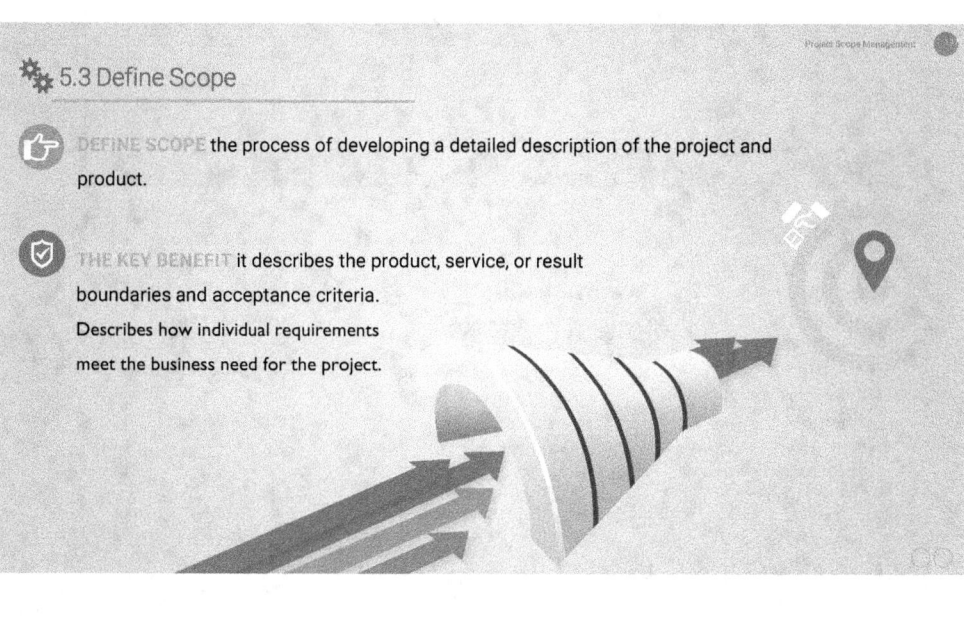

## 5.3 Define Scope

| Inputs | Inputs Tools & Techniques Outputs | Outputs |
|---|---|---|
| 1 Project charter<br>2 Project management plan<br>　• Scope management plan<br>3 Project documents<br>　• Assumption log<br>　• Requirements documentation<br>　• Risk register<br>4 Enterprise environmental factors<br>5 Organizational process assets | .1 Expert judgment<br>.2 Data analysis<br>　• Alternatives analysis<br>.3 Decision making<br>　• Multicriteria decision analysis<br>.4 Interpersonal and team skills<br>　• Facilitation<br>.5 Product analysis | 1 Project scope statement<br>2 Project documents updates<br>　• Assumption log<br>　• Requirements documentation<br>　• Requirements traceability matrix<br>　• Stakeholder register |

## 5.3 Define Scope  Input

**01 PROJECT CHARTER**

**02 PROJECT MANAGEMENT PLAN**
- Assumption log: identifies assumptions and constraints about the product, project, environment, stakeholders, and other factors that can influence the project
- Requirements documentation.
- Risk register: response strategies that may affect the scope, to avoid or mitigate a risk.

**03 ENTERPRISE ENVIRONMENTAL FACTORS**

**04 ORGANIZATIONAL PROCESS ASSETS**

## 5.3 Define Scope  Tools & Techniques

**01 EXPERT JUDGMENT**

**02 DATA ANALYSIS**
Alternatives analysis can be used to evaluate ways to meet the requirements and the objectives identified in the charter.

**03 DECISION MAKING**

**04 INTERPERSONAL AND TEAM SKILLS**

**05 PRODUCT ANALYSIS**
- Product breakdown,
- Requirements analysis,
- Systems analysis,
- Systems engineering,
- Value analysis, and
- Value engineering.

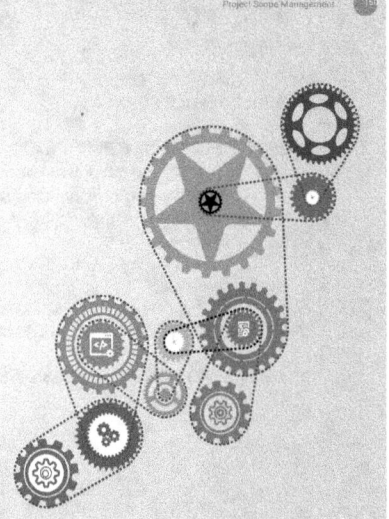

## 5.3 Define Scope    Output

**① PROJECT SCOPE STATEMENT**
- ➤ Description of the project scope, major deliverables, assumptions, and constraints.
- ➤ project scope statement contains a detailed description of the scope components. These components are progressively elaborated throughout the project
- ➤ The detailed project scope statement includes:
  - Product scope description.
  - Deliverables.
  - Acceptance criteria.
  - Project exclusions.

**② PROJECT DOCUMENTS UPDATES**
- Assumption log.
- Requirements documentation.
- Requirements traceability matrix.
- Stakeholder register.

## 5.4 Create WBS

 CREATE WBS IS the process of subdividing project deliverables and project work into smaller, more manageable components.

 THE KEY BENEFIT it provides a framework of what has to be delivered.

WBS is a hierarchical decomposition of the total scope of work. specified in the current approved project scope statement.

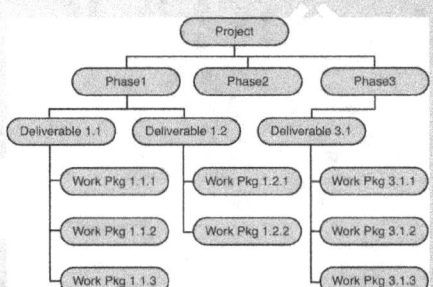

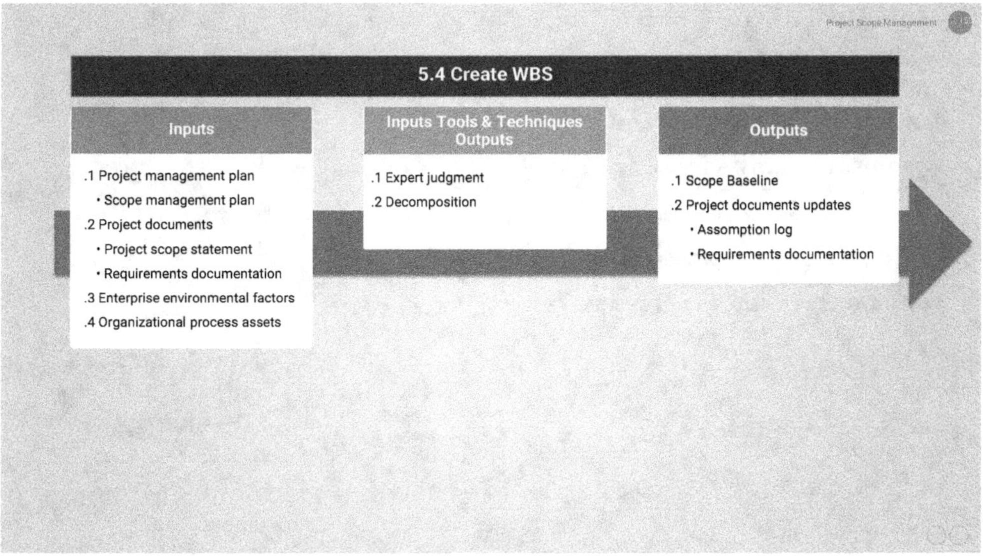

## 5.4 Create WBS

**Inputs**
- .1 Project management plan
  - Scope management plan
- .2 Project documents
  - Project scope statement
  - Requirements documentation
- .3 Enterprise environmental factors
- .4 Organizational process assets

**Inputs Tools & Techniques Outputs**
- .1 Expert judgment
- .2 Decomposition

**Outputs**
- .1 Scope Baseline
- .2 Project documents updates
  - Assomption log
  - Requirements documentation

## 5.4 Create WBS — Input

**01 PROJECT MANAGEMENT PLAN**

**02 PROJECT DOCUMENTS**
- Project scope statement.
- Requirements documentation.

**03 ENTERPRISE ENVIRONMENTAL FACTORS (EEF)**

**04 ORGANIZATIONAL PROCESS ASSETS (OPA)**

## 5.4 Create WBS  Tools & Techniques

 **EXPERT JUDGMENT**

 **DECOMPOSITION**
Decomposition is a technique used for dividing and subdividing the project scope and project deliverables into smaller, more manageable parts.

The work package is the work defined at the lowest level of the WBS for which cost and duration can be estimated and managed.

- Identifying and analyzing the deliverables and related work,
- Structuring and organizing the WBS,
- Decomposing the upper WBS levels into lower-level detailed components,
- Developing and assigning identification codes to the WBS components,
- Verifying that the degree of decomposition of the deliverables is appropriate

## 5.4 Create WBS  Output

**SCOPE BASELINE**
- Project scope statement.
- WBS.
- Work package.
  - The lowest level of the WBS with a unique identifier. And part of a control account.
  - These identifiers provide a structure for hierarchical summation of costs, schedule, and resource information and form a code of accounts.
  - control account is a management control point where scope, budget, and schedule are integrated and compared to the earned value for performance measurement.
  - control account has two or more work packages, though each work package is associated with a single control account.

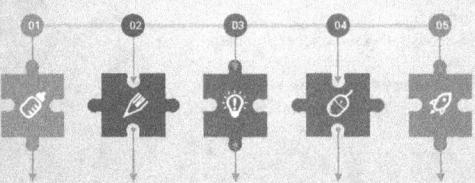

## 5.4 Create WBS  Output

- **Planning package.** is a work breakdown structure component below the control account and above the work package with known work

- **WBS dictionary** is a document that provides detailed deliverable, activity, and scheduling information about each component in the WBS.

  - Code of account identifier,
  - Description of work,
  - Assumptions and constraints,
  - Responsible organization,
  - Schedule milestones,
  - Associated schedule activities,

  - Resources required,
  - Cost estimates,
  - Quality requirements,
  - Acceptance criteria,
  - Technical references, and
  - Agreement information.

**PROJECT DOCUMENTS UPDATES**
- Assumption log.
- Requirements documentation.

## 5.5 Validate Scope

 **VALIDATE SCOPE** the process of formalizing acceptance of the completed project deliverables.

 **THE KEY BENEFIT** it brings objectivity to the acceptance process and increases the probability of final product, service, or result acceptance by validating each deliverable.

## 5.5 Validate Scope

### Inputs

.1 Project management plan
  - Scope management plan
  - Requirements management plan
  - Scope baseline
.2 Project documents
  - Lessons learned register
  - Quality reports
  - Requirements documentation
  - Requirements traceability matrix
.3 Verified deliverables
.4 Work performance data

### Inputs Tools & Techniques Outputs

.1 Inspection
.2 Decision making
  - Voting

### Outputs

.1 Accepted deliverables
.2 Work performance information
.3 Change requests
.4 Project document updates
  - Lessons learned register
  - Requirements documentation
  - Requirements traceability matrix

## 5.5 Validate Scope — Input

**01 Project management plan**
- Scope management plan
- Requirements management plan
- Scope baseline

**02 Project documents**
- Lessons learned register
- Quality reports
- Requirements documentation
- Requirements traceability matrix

**03 Verified deliverables**

**04 Work performance data**

## 5.5 Validate Scope — Tools & Techniques

**01 INSPECTION**
- includes activities such as measuring, examining, and validating to
- determine whether work and deliverables meet requirements and product acceptance criteria.
- Inspections are sometimes called reviews, product reviews, and walkthroughs.

**02 DECISION MAKING**
Voting is used to reach a conclusion when the validation is performed by the project team and other stakeholders.

## 5.5 Validate Scope — Output

**01 ACCEPTED DELIVERABLES**
Deliverables that meet the acceptance criteria are formally signed off and approved by the customer or sponsor.

**02 WORK PERFORMANCE INFORMATION**
Work performance information includes information about project progress, such as which deliverables have been accepted and which have not been accepted and the reasons why.

**03 CHANGE REQUESTS**
The completed deliverables that have not been formally accepted are documented, along with the reasons for non-acceptance of those deliverables.
Those deliverables may require a change request for defect repair.

**04 PROJECT DOCUMENTS UPDATES**
- Lessons learned register.
- Requirements documentation.
- Requirements traceability matrix.

## 5.6 Control Scope

 **CONTROL SCOPE** is the process of monitoring the status of the project and product scope and managing changes to the scope baseline.

 **THE KEY BENEFIT** the scope baseline is maintained throughout the project.

> Controlling the project scope ensures all requested changes and recommended corrective or preventive actions are processed through the Perform Integrated Change Control process.

> The uncontrolled expansion to product or project scope without adjustments to time, cost, and resources is referred to as scope creep.

# 5.6 Control Scope

## Inputs

1 Project management plan
- Scope management plan
- Requirements management plan
- Change management plan
- Configuration management plan
- Scope baseline
- Performance measurement baseline

.2 Project documents
- Lessons learned register
- Requirements documentation
- Requirements traceability matrix

.3 Work performance data
.4 Organizational process assets

## Inputs Tools & Techniques Outputs

.1 Data analysis
- Variance analysis
- Trend analysis

## Outputs

.1 Work performance information
.2 Change requests
.3 Project management plan updates
- Scope management plan
- Scope baseline
- Schedule baseline
- Cost baseline
- Performance measurement baseline

.4 Project documents updates
- Lessons learned register
- Requirements documentation
- Requirements traceability matrix

## 5.6 Control Scope — Input

**01 PROJECT MANAGEMENT PLAN**
- Scope management plan.
- Requirements management plan.
- Change management plan.
- Configuration management plan.
- Scope baseline.
- Performance measurement baseline.

**02 PROJECT DOCUMENTS**
- Lessons learned register
- Requirements documentation.
- Requirements traceability matrix.

**03 WORK PERFORMANCE DATA**

**04 ORGANIZATIONAL PROCESS ASSETS**

## 5.6 Control Scope  Tools & Techniques

 DATA ANALYSIS
- **Variance analysis.**
is used to compare the baseline to the actual results and determine if the variance is within the threshold amount or if corrective or preventive action is appropriate.

- **Trend analysis.**
examines project performance over time to determine if performance is improving or deteriorating

## 5.6 Control Scope — Output

**01 WORK PERFORMANCE INFORMATION**

**02 CHANGE REQUESTS**

**03 PROJECT MANAGEMENT PLAN UPDATES**
- Scope management plan.
- Scope baseline.
- Schedule baseline.
- Cost baseline.
- Performance measurement baseline.

**04 PROJECT DOCUMENTS UPDATES**
- Lessons learned register.
- Requirements documentation.
- Requirements traceability matrix.

## Have You Any question?

| What | Why | When | Where | How | who |

# 6. PROJECT
## SCHEDULE MANAGEMENT

Presented by :
**Nasser Al Mohimeed**
PMO Director, ISO 21500 Lead Project Manager
Certified Project Managers Trainer

# 7. PROJECT
## COST MANAGEMENT

Presented by :
**Nasser Al Mohimeed**
PMO Director, ISO 21500 Lead Project Manager
Certified Project Managers Trainer

## Project Cost Management

Project Cost Management includes the processes involved in planning, estimating, budgeting, financing, funding, managing, and controlling costs so that the project can be completed within the approved budget

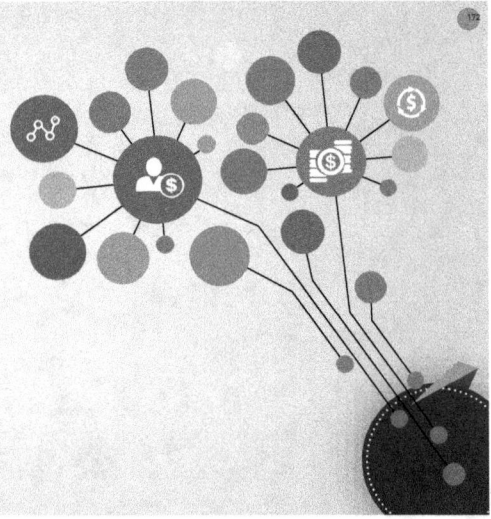

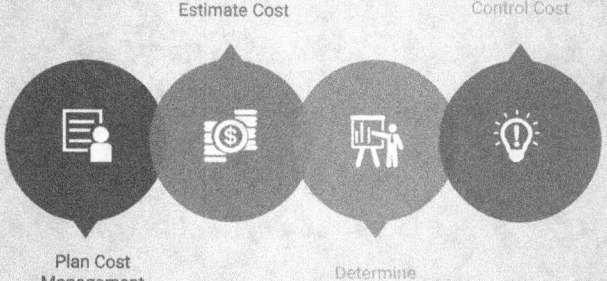

# Project Cost Management

| Initiating | Planning | Executing | Monitoring & Controlling | Closing |
|---|---|---|---|---|
| | 7.1 Plan Cost Management | | 7.4 Control Cost | |
| | 7.2 Estimate Cost | | | |
| | 7.3 Determine Budget | | | |

### Key concepts for Project Cost Management

- Project Cost Management is primarily concerned with the cost of the resources needed to complete project activities.

- Project Cost Management should consider the effect of project decisions on the subsequent recurring cost of using, maintaining, and supporting the product, service or result of the project.

- Cost management is recognizing that different stakeholders measure project costs in different ways and at different times.

- Predicting and analyzing the prospective financial performance of the project's may be performed outside of the project.

- When predictions and analyses are included in the project, Project Cost Management may address additional processes and numerous general financial management techniques such as return on investment, discounted cash flow, and investment payback analysis.

## TRENDS AND EMERGING PRACTICES IN PROJECT COST MANAGEMENT

- include the expansion of earned value management (EVM) to include the concept of earned schedule (ES).

- Earned schedule theory replaces the schedule variance measures used in traditional EVM (earned value – planned value) with ES and actual time (AT). Using the alternate equation for calculating schedule variance ES – AT,

- if the amount of earned schedule is greater than 0, then the project is considered ahead of schedule.

- The schedule performance index (SPI) using earned schedule metrics is ES/AT. This indicates the efficiency with which work is being accomplished.

- Earned schedule theory also provides formulas for forecasting the project completion date, using earned schedule, actual time, and estimated duration

Key concepts for Project Cost Management

**TAILORING CONSIDERATIONS**

Project Cost Management processes are applied. Considerations for tailoring include :

- **Knowledge management.** Does the organization have a formal knowledge management and financial database repository that a project manager is required to use and that is readily accessible?
- **Estimating and budgeting.** Does the organization have existing formal or informal cost estimating and budgeting-related policies, procedures, and guidelines?
- **Earned value management.** Does the organization use earned value management in managing projects?
- **Use of agile approach.** Does the organization use agile methodologies in managing projects? How does this impact cost estimating?
- **Governance.** Does the organization have formal or informal audit and governance policies, procedures, and guidelines?

## Key concepts for Project Cost Management

**CONSIDERATIONS FOR AGILE/ADAPTIVE ENVIRONMENTS**

- detailed cost calculations is not benefit to the Projects with high degrees of uncertainty due to frequent changes.
- lightweight estimation methods can be used to generate a fast, high-level forecast of project labor costs, which can then be easily adjusted as changes arise.
- Detailed estimates are reserved for short-term planning horizons in a just-in-time fashion.
- In cases where high-variability projects are also subject to strict budgets, the scope and schedule are more often adjusted to stay within cost constraints.

## 7.1 Plan Cost Management

Is the process of defining how the project costs will be estimated, budgeted, managed, monitored, and controlled

It provides guidance and direction on how the project costs will be managed throughout the project

Plan Cost Management

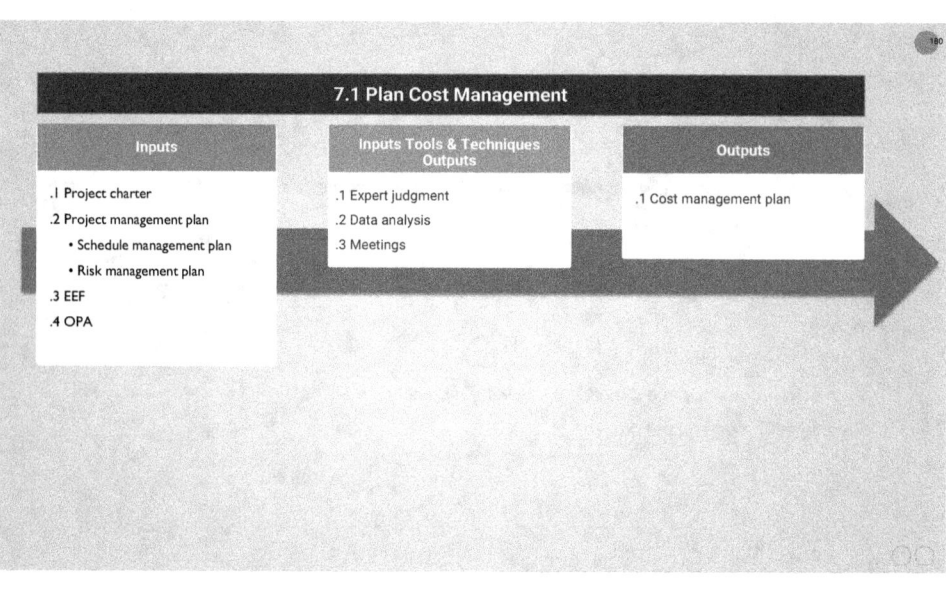

## 7.1 Plan Cost Management — Input

**01 PROJECT CHARTER**

**02 PROJECT MANAGEMENT PLAN**
- Schedule management plan
- Risk management plan

**03 ENTERPRISE ENVIRONMENTAL FACTORS**

**04 ORGANIZATIONAL PROCESS ASSETS**

## 7.1 Plan Cost Management — Tools & Techniques

01 Expert judgment

02 Data Analysis
- Alternatives analysis

03 Meetings

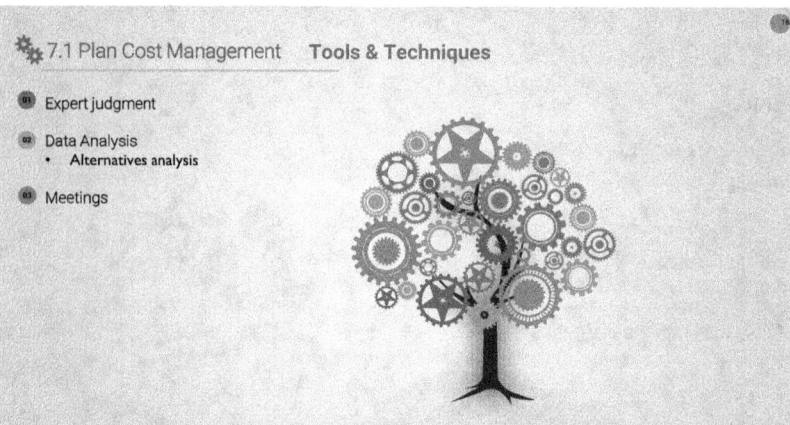

## 7.1 Plan Cost Management — Output 1

**Cost management plan**
Describes how the project costs will be planned, structured, and controlled. The cost management processes and their associated tools and techniques are documented in the cost management plan.

- **Units of measure.** Each unit used in measurements is defined for each of the resources.
- **Level of precision.** This is the degree to which cost estimates will be rounded up or down (e.g., US$995.59 to US$1,000
- **Level of accuracy.** The acceptable range (e.g., ±10%) used in determining realistic cost estimates is specified, and may include an amount for contingencies.
- **Organizational procedures links.** Each control account is assigned a unique code or account number(s) that links directly to the performing organization's accounting system.

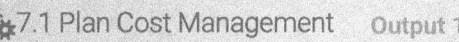

## 7.1 Plan Cost Management — Output 2

**Cost management plan (continue)**
- Control thresholds. the percentage deviations from the baseline plan.
- Rules of performance measurement. EVM rules
  - The points in the WBS at which measurement of control accounts will be performed;
  - Establish the EVM techniques (e.g., weighted milestones, fixed-formula, percent complete, etc.) to be employed;
  - Tracking methodologies and the EVM computation equations for calculating (EAC) forecasts
- Reporting formats. The formats and frequency for cost reports
- Additional details.
  - Description of strategic funding choices,
  - Procedure to account for fluctuations in currency exchange rates,
  - Procedure for project cost recording.

## 7.2 Estimate Costs

**ESTIMATE COSTS**
s the process of developing an approximation of the cost of resources needed to complete project work

**THE KEY BENEFIT**
Determines the monetary resources required for the project

## 7.2 Estimate Costs

### Inputs
.1 Project management plan
  - Cost management plan
  - Quality management plan
  - Scope baseline
.2 Project documents
  - Lessons learned register
  - Project schedule
  - Resources requirements
  - Risk register
.3 EEF
.4 OPA

### Tools & Techniques
.1 Expert judgment
.2 Analogous estimating
.3 Parametric estimating
.4 Bottom-up estimating
.5 Three-point estimating
.6 Data analysis
  - Alternatives analysis
  - Reserve analysis
  - Cost of quality
.7 PMIS
.8 Decision making

### Outputs
.1 Cost estimates
.2 Basis of estimates
.3 Project documents updates
  - Assumption log
  - Lessons learned register
  - Risk register

## 7.2 Estimate Costs  Input

**01 PROJECT MANAGEMENT PLAN**
- Cost management plan
- Quality management plan
- Scope baseline

**02 PROJECT DOCUMENTS**
- Lessons learned register
- Project schedule
- Resources requirements
- Risk register

**03 ENTERPRISE ENVIRONMENTAL FACTORS**

**04 ORGANIZATIONAL PROCESS ASSETS**

## 7.2 Estimate Costs — Tools & Techniques 1

- **01 EXPERT JUDGMENT**
- **02 ANALOGOUS ESTIMATING**
- **03 PARAMETRIC ESTIMATING**
- **04 BOTTOM-UP ESTIMATING**
- **05 THREE-POINT ESTIMATING**
  - Triangular distribution. $cE = (cO + cM + cP) / 3$
  - Beta distribution. $cE = (cO + 4cM + cP) / 6$
- **06 PMIS**
- **07 Decision making**

## 7.2 Estimate Costs — Tools & Techniques 2

**DATA ANALYSIS**

- ❖ **Alternatives analysis:** evaluate identified options select which options to use to execute the work
  e.g. evaluating the cost, schedule, resource, and quality impacts of buying versus making

- ❖ **Reserve analysis.**
  - Cost estimates may include contingency reserves to account for cost uncertainty.
  - Contingency reserves are often viewed as the part of the budget intended to address the known-unknowns that can affect a project. e.g. rework for some project deliverables
  - Contingency reserves can be provided at any level from the specific activity to the entire project.
  - The contingency reserve may be a percentage of the estimated cost, a fixed number, or may be developed by usin quantitative analysis methods.
  - Contingency should be clearly identified in cost documentation.
  - Contingency reserves are part of the cost baseline and the overall funding requirements for the project.

- ❖ **Cost of quality.** Includes evaluating the cost impact of additional investment in conformance versus the cost of nonconformance.

## 7.2 Estimate Costs  Output 1

**COST ESTIMATES**
Include quantitative assessments of the costs required to complete project work, as well as contingency amounts to account for identified risks, and management reserve to cover unplanned work.
- <u>direct cost</u> (labor, materials, equipment, services, facilities, information technology, and special categories such as cost of financing (including interest charges), an inflation allowance, exchange rates, or a cost contingency reserve.
- <u>Indirect costs</u>, can be included at the activity level or at higher levels.

**BASIS OF ESTIMATES**
should provide a clear and complete understanding of how the cost estimate was derived.
- Documentation of the basis of the estimate (i.e., how it was developed),
- Documentation of all assumptions made,
- Documentation of any known constraints,
- Documentation of identified risks included when estimating costs,
- Indication of the range of possible estimates (e.g., US$10,000 (±10%) to indicate that the item is expected to cost between a range of values),
- Indication of the confidence level of the final estimate.

## 7.2 Estimate Costs — Output 2

**PROJECT DOCUMENTS UPDATES**

- Assumption log
- Lessons learned register
- Risk register

## 7.3 Determine Budget

**DETERMINE BUDGET**
aggregating the estimated costs of individual activities or work packages to establish an authorized cost baseline.

**THE KEY BENEFIT**
determines the cost baseline against which project performance can be monitored and controlled

 A project budget includes all the funds authorized to execute the project. The cost baseline is the approved version of the time-phased project budget that includes contingency reserves, but excludes management reserves.

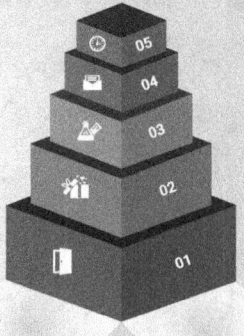

## 7.3 Determine Budget

**Inputs**

.1 Project management plan
  - Cost management plan
  - Resource management plan
  - Scope baseline
.2 Project documents
  - Basis of estimates
  - Cost estimates
  - Project schedule
  - Risk register
.3 Business documents
  - Business case
  - Benefits management plan
.4 Agreements
.5 EEF
.6 OPA

**Inputs Tools & Techniques Outputs**

.1 Expert judgment
.2 Cost aggregation
.3 Data analysis
  - Reserve analysis
.4 Historical information review
.5 Funding limit reconciliation
.6 Financing

**Outputs**

.1 Cost baseline
.2 Project funding requirements
.3 Project documents updates
  - Cost estimates
  - Project schedule
  - Risk register

## 7.3 Determine Budget — Input

**01 Project management plan**
- Cost management plan
- Resource management plan
- Scope baseline

**02 Project documents**
- Basis of estimates
- Cost estimates
- Project schedule
- Risk register

**03 Business documents**
- Business case
- Benefits management plan

## 7.3 Determine Budget  Input

- **04** Agreements
- **05** Enterprise environmental factors
- **06** Organizational process assets

## 7.3 Determine Budget — Tools & Techniques 1

**01 EXPERT JUDGMENT**

**02 COST AGGREGATION**
The work package cost estimates are then aggregated for the higher component levels of the WBS (such as control accounts) and, ultimately, for the entire project.

**03 DATA ANALYSIS**
- Reserve analysis

**04 HISTORICAL INFORMATION REVIEW**
Assist in developing parametric estimates or analogous estimates.

**05 FUNDING LIMIT RECONCILIATION**
The expenditure of funds should be reconciled with any funding limits on the commitment of funds for the project.

**06 FINANCING**

## Project budget.

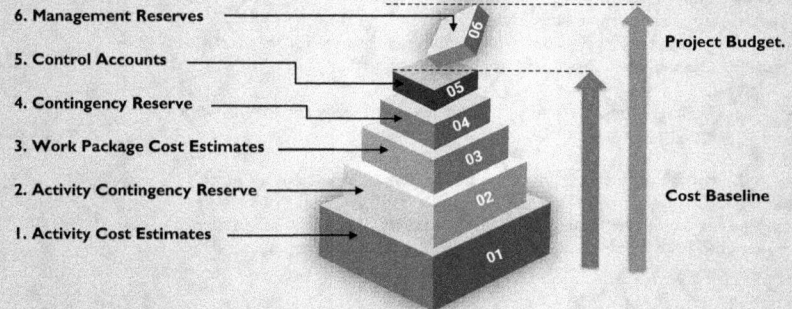

- 6. Management Reserves
- 5. Control Accounts
- 4. Contingency Reserve
- 3. Work Package Cost Estimates
- 2. Activity Contingency Reserve
- 1. Activity Cost Estimates

Project Budget.

Cost Baseline

## 7.3 Determine Budget  Output 1

**COST BASELINE**

It is the approved version of the project budget, excluding any management reserves,
Can only be changed through formal change control procedures.
It is used as a basis for comparison to actual results.

1. Cost estimates for the project activities, with contingency reserves for these activities, are aggregated into their associated work package costs.

2. The work package cost estimates, with contingency reserves estimated for the work packages, are aggregated into control accounts.

3. The summation of the control accounts make up the cost baseline.

4. Cost baseline are directly tied to the schedule activities; For projects that use earned value management, the cost baseline is referred to as the performance measurement baseline.

5. Management reserves are added to the cost baseline to produce the project budget.

## Project budget.

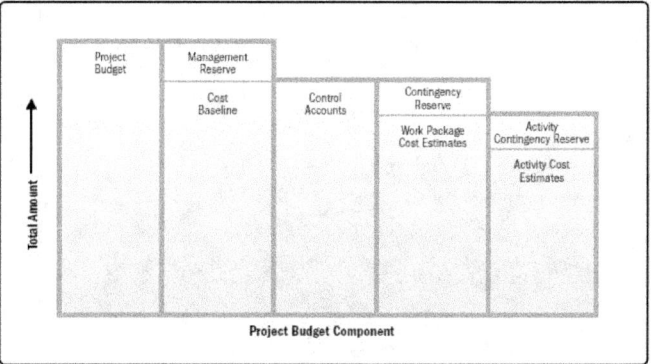

## Cost Baseline, Expenditures, and Funding Requirements

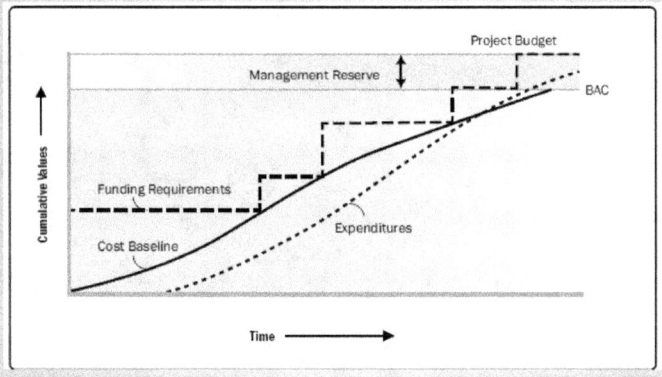

## 7.3 Determine Budget — Output 2

**PROJECT FUNDING REQUIREMENTS**
- Total funding requirements and periodic funding requirements (e.g., quarterly, annually) are derived from the cost baseline.
- The cost baseline will include projected expenditures plus anticipated liabilities. Funding often occurs in incremental amounts, and may not be evenly distributed,
- Total funds required are those included in the cost baseline plus management reserves

**PROJECT DOCUMENTS UPDATES**
- Cost estimates.
- Project schedule
- Risk register

## 7.4 Control Costs

**CONTROL COSTS**

is the process of monitoring the status of the project to update the project costs and managing changes to the cost baseline

**THE KEY BENEFIT**

is that the cost baseline is maintained throughout the project

## 7.4 Control Costs

**Inputs**
- .1 Project management plan
  - Cost management plan
  - Cost baseline
  - Performance measurement baseline
- .2 Project documents
  - Lessons learned register
- .3 Project funding requirements
- .4 EEF
- .5 OPA

**Tools & Techniques**
- .1 Expert judgment
- .2 Data analysis
  - Earned value analysis
  - Variance analysis
  - Trend analysis
  - Reserve analysis
- .3 To-complete performance index
- .4 PMIS

**Outputs**
- .1 Work performance information
- .2 Cost forecasts
- .3 Change requests
- .4 Project management plan updates
  - Cost management plan
  - Cost baseline
  - Performance measurement baseline
- .5 Project documents updates
  - Assumption log
  - Basis of estimates
  - Cost estimates
  - Lessons learned register
  - Risk register

## 7.4 Control Costs　Input

**01　Project management plan**
- Cost management plan
- Cost baseline
- Performance measurement baseline

**02　Project documents**
- Lessons learned register

**03　Project funding requirements**

**04　Enterprise environmental factors**

**05　Organizational process assets**

## 7.4 Control Costs — Tools & Techniques 1

**① EXPERT JUDGMENT**

**② DATA ANALYSIS**

**Earned value analysis (EVA).** compares the performance measurement baseline to the actual schedule and cost performance. develops and monitors three key dimensions for each work package and control account:

- ❖ Planned value (PV) is the authorized budget assigned to scheduled work. (WBS)
- planned value defines the physical work that should have been accomplished.
- The total PV referred to as the performance measurement baseline (PMB).
- The total PV for the project is also known as budget at completion (BAC).

- ❖ Earned value (EV) It is the budget associated with the authorized work that has been completed.
- EV measured cannot be greater than the authorized PV budget for a component.

- ❖ Actual cost (AC) the total cost incurred in accomplishing the work that the EV measured
- The AC will have no upper limit; whatever is spent to achieve the EV will be measured.

## 7.4 Control Costs — Tools & Techniques 2

**Variance analysis.**
Performed by comparing planned cost against actual cost to identify variances between the cost baseline and actual project performance.
Determining the cause and degree of variance relative to the cost baseline and deciding whether corrective or preventive action is required.

- Schedule variance (SV) is a measure of schedule performance as difference between the earned value and the planned value.
  it can indicate when a project is falling behind or is ahead of its baseline schedule.
  $SV = EV - PV$
  Positive = Ahead of Schedule
  Neutral = On schedule
  Negative = Behind Schedule

- Cost variance (CV) is the amount of budget deficit or surplus at a given point in time, expressed as the difference between earned value and the actual cost.
  $CV = EV - AC$
  Positive = Under planned cost
  Neutral = On planned cost
  Negative = Over planned cost

## 7.4 Control Costs — Tools & Techniques 3

- The cost variance at the end of the project will be the difference between the budget at completion (BAC) and the actual amount spent.
  VAC = BAC – EAC

- Schedule performance index (SPI) is a measure of schedule efficiency expressed as the ratio of earned value to planned value. It measures how efficiently the project team is accomplishing the work.
  SPI = EV/PV.
  SPI < 1.0 indicates less work was completed than was planned.
  SPI > 1.0 indicates that more work was completed than was planned.
  SPI = Neutral indicates On schedule

- Cost performance index (CPI) is a measure of the cost efficiency of budgeted resources,
  CPI = EV/AC
  CPI < 1.0 indicates a cost overrun
  CPI > 1.0 indicates a cost underrun of performance to date.
  CPI= Neutral indicates a cost On planned cost

## 7.4 Control Costs — Tools & Techniques 4

**Trend analysis** examines project performance over time to determine if performance is improving or deteriorating.

- **Charts**. PV, EV, AC can be monitored and reported periodically   e.g. S-curves

- **Forecasting**. the project team may develop a forecast (EAC) that may differ from the budget at completion (BAC) based on the project performance

    1. If the CPI is expected to be the same for the remainder of the project,
       EAC = BAC/CPI

    2. If future work will be accomplished at the planned rate,
       EAC = AC + BAC − EV

    3. If the initial plan is no longer valid,
       EAC = AC + Bottom-up ETC

    4. If both the CPI and SPI influence the remaining work,
       EAC = AC + [(BAC − EV)/(CPI × SPI)]

## S-curves: PV, EV, AC monitored and reported periodically

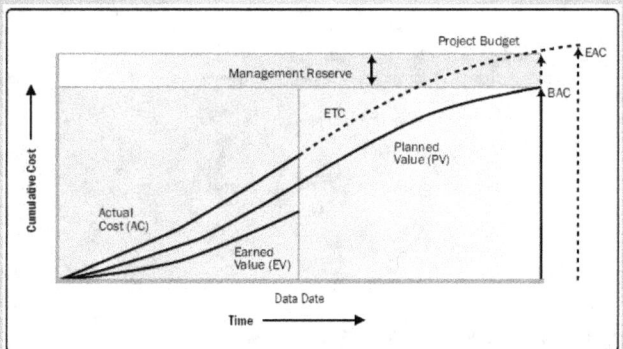

## 7.4 Control Costs — Tools & Techniques 5

**Reserve analysis**

 **TO-COMPLETE PERFORMANCE INDEX**

A measure of the cost performance that must be achieved with the remaining resources in order to meet a specified management goal.
- The efficiency that must be maintained in order to complete on plan.
  TCPI = (BAC − EV)/(BAC − AC)

- The efficiency that must be maintained in order to complete the current EAC.
  TCPI = (BAC − EV)/(EAC − AC)

  TCPI > 1.0   Harder to complete
  TCPI = 1.0   Same to complete
  TCPI < 1.0   Easier to complete

**PROJECT MANAGEMENT INFORMATION SYSTEM (PMIS)**

## 7.4 Control Costs  Output

**01** **Work performance information**

**02** **Cost forecasts**

**03** **Change requests**

**04** **Project management plan updates**
- Cost management plan
- Cost baseline
- Performance measurement baseline

**05** **Project documents updates**
- Assumption log
- Basis of estimates
- Cost estimates
- Lessons learned register
- Risk register

## Have You Any question?

| What | Why | When | Where | How | who |

212

# 8. PROJECT
## QUALITY MANAGEMENT

Presented by :
**Nasser Al Mohimeed**
PMO Director, ISO 21500 Lead Project Manager
Certified Project Managers Trainer

## Project Quality Management

Project Quality Management includes the processes for incorporating the organization's quality policy regarding planning, managing, and controlling project and product quality requirements in order to meet stakeholders' objectives.

Project Quality Management also supports continuous process improvement activities as undertaken on behalf of the performing organization.

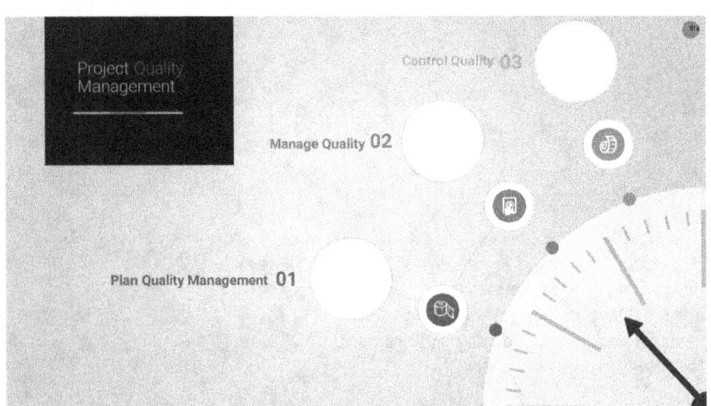

## Project Quality Management

| Initiating | Planning | Executing | Monitoring & Controlling | Closing |
|---|---|---|---|---|
| | 8.1 Plan Quality Management | 8.2 Manage Quality | 8.3 Control Quality | |

## Quality Management Process Interrelations

- The Plan Quality Management process is concerned with the quality that the work needs to have.

- Manage Quality is concerned with managing the quality processes throughout the project.

- Control Quality is concerned with comparing the work results with the quality requirements to ensure the result is acceptable.

- There are two outputs specific to the Project Quality Management Knowledge Area that are used by other Knowledge Areas:
  - verified deliverables
  - quality reports.

## Key concepts for Project Quality Management

- **Project Quality Management** addresses the management of the project and it's deliverables. It applies to all projects, regardless of the nature of their deliverables.

- **Quality measures and techniques** are specific to the type of deliverables
  E.g. software deliverables use different approaches and measures from those used when building a nuclear power plant

- **Quality Vs grade.**

  **Quality** the degree to which a set of inherent characteristics fulfill requirements (ISO 9000).

  **Grade** is a category assigned to deliverables having the same functional use but different technical characteristics.

  - PM and PM team are responsible for delivering the required levels of both quality and grade.
  - Low quality level is always a problem, a low-grade product may not be a problem.

    E.g. A low-grade product (with a limited features) is of high quality (no defects). the product would be appropriate for its general purpose of use.

## Key concepts for Project Quality Management

- **Prevention Vs inspection.** It is better to design quality into deliverables, rather than to find quality issues during inspection.

  The cost of preventing mistakes is generally much less than the cost of correcting mistakes

  - **Prevention** (keeping errors out of the process) and **inspection** (keeping errors out of the hands of the customer);
  - **Attribute sampling** (the result either conforms or does not conform) and **variable sampling** (the result is rated on a continuous scale that measures the degree of conformity);
  - **Tolerances** (specified range of acceptable results) and **control limits** (that identify the boundaries of common variation in a statistically stable process or process performance).

- **Cost of quality (COQ)** includes all costs incurred over the life of the product by investment in preventing nonconformance to requirements, appraising the product or service for conformance to requirements,

  **Cost of poor quality** failing to meet requirements (rework). categorized into internal (found by the project team) and external (found by the customer).

  COQ is concern of PM, program management, portfolio management, PMO, or operations management

## Key concepts for Project Quality Management

▇ **There are five levels of increasingly effective quality management :**
- Most expensive approach is to let the customer find the defects. This approach can lead to warranty issues, recalls, loss of reputation, and rework costs.
- Detect and correct the defects before the deliverables are sent to the customer as part of the quality control process. The control quality process has related costs, which are mainly the appraisal costs and internal failure costs.
- Use quality assurance to examine and correct the process itself and not just special defects.
- Incorporate quality into the planning and designing of the project and product.
- Create a culture throughout the organization that is aware and committed to quality in processes and products.

## TRENDS AND EMERGING PRACTICES IN PROJECT QUALITY MANAGEMENT

Modern quality management approaches seek to minimize variation and to deliver results that meet defined stakeholder requirements

- **Customer satisfaction.** Understand, evaluate, define, and manage requirements so that customer expectations are met. This requires a combination of
  - conformance to requirements (to ensure the project produces what it was created to produce) and
  - fitness for use (the product or service needs to satisfy the real needs).
  - ➤ In agile environments, stakeholder engagement with the team ensures customer satisfaction is maintained throughout the project.

- **Continual improvement.** improve both the quality of project management, as well as the quality of the end product, service, or result.
  - The plan-do-check-act (PDCA) defined by Shewhart and modified by Deming.
  - Total quality management (TQM),
  - Six Sigma, and Lean Six Sigma

## TRENDS AND EMERGING PRACTICES IN PROJECT QUALITY MANAGEMENT

**Management responsibility.**
Management retains, within its responsibility for quality, a related responsibility to provide suitable resources at adequate capacities.

**Mutually beneficial partnership with suppliers**
- Relationships based on partnership and cooperation with the supplier are more beneficial to the organization and to the suppliers than traditional supplier management.
- The organization should prefer long-term relationships over short-term gains.
- A mutually beneficial relationship enhances the ability for both the organization and the suppliers to create value for each other,

## TAILORING CONSIDERATIONS

Each project is unique; therefore, the project manager will need to tailor the way Project Quality Management processes are applied

- **Policy compliance and auditing.** What quality policies and procedures exist in the organization? What quality tools, techniques, and templates are used in the organization?
- **Standards and regulatory compliance.** Are there any specific quality standards in the industry that need to be applied? Are there any specific governmental, legal, or regulatory constraints that need to be taken into consideration?
- **Continuous improvement.** How will quality improvement be managed in the project? Is it managed at the organizational level or at the level of each project?
- **Stakeholder engagement.** Is there a collaborative environment for stakeholders and suppliers?

## CONSIDERATIONS FOR AGILE/ADAPTIVE ENVIRONMENTS

- Agile methods call for frequent quality and review steps built in throughout the project rather than toward the end of the project.

- Recurring retrospectives regularly check on the effectiveness of the quality processes. They look for the root cause of issues then suggest trials of new approaches to improve quality.

- Subsequent retrospectives evaluate any trial processes to determine if they are working and should be continued or new adjusting or should be dropped from use.

- In order to facilitate frequent, incremental delivery, agile methods focus on small batches of work, incorporating as many elements of project deliverables as possible.

- Small batch systems aim to uncover inconsistencies and quality issues earlier in the project life cycle when the overall costs of change are lower.

## 8.1 Plan Quality Management

**PLAN QUALITY MANAGEMENT**

Is the process of identifying quality requirements and/or standards for the project and its deliverables, and documenting how the project will demonstrate compliance with quality requirements and/or standards.

**THE KEY BENEFIT**

IT provides guidance and direction on how quality will be managed and verified throughout the project

## 8.1 Plan Quality Management

### Inputs

.1 Project charter
.2 Project management plan
  • Requirements management plan
  • Risk management plan
  • Stakeholder engagement plan
  • Scope baseline
.3 Project documents
  • Assumption log
  • Requirements documentation
  • Requirements traceability matrix
  • Risk register
  • Stakeholder register
.4 EEF
.5 OPA

### Inputs Tools & Techniques Outputs

.1 Expert judgment
.2 Data gathering
  • Benchmarking
  • Brainstorming
  • Interviews
.3 Data analysis
  • Cost-benefit analysis
  • Cost of quality
.4 Decision making
  • Multicriteria decision analysis
.5 Data representation
  • Flowcharts
  • Logical data model
  • Matrix diagrams
  • Mind mapping
.6 Test and inspection planning
.7 Meetings

### Outputs

.1 Quality management plan
.2 Quality metrics
.3 Project management plan updates
  • Risk management plan
  • Scope baseline
.4 Project documents updates
  • Lessons learned register
  • Requirements traceability matrix
  • Risk register
  • Stakeholder register

## 8.1 Plan Quality Management 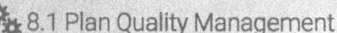 Input

**01 PROJECT CHARTER**

**02 PROJECT MANAGEMENT PLAN**
- Requirements management plan
- Risk management plan
- Stakeholder engagement plan
- Scope baseline

**03 PROJECT DOCUMENTS**
- Assumption log
- Requirements documentation
- Requirements traceability matrix
- Risk register
- Stakeholder register

**04 ENTERPRISE ENVIRONMENTAL FACTORS**

**05 ORGANIZATIONAL PROCESS ASSETS**

## 8.1 Plan Quality Management — Tools & Techniques 1

**01 EXPERT JUDGMENT**

**02 DATA GATHERING**
- Benchmarking.
- Brainstorming.
- Interviews.

**03 DATA ANALYSIS**
Cost-benefit analysis
- Is a financial analysis tool used to estimate the strengths and weaknesses of alternatives in order to determine the best alternative in terms of benefits provided.
- Help PM to determine if the planned quality activities are cost effective.
- The primary benefits of meeting quality requirements include less rework, higher productivity, lower costs, increased stakeholder satisfaction, and increased profitability.
- A cost-benefit analysis for each quality activity compares the cost of the quality step to the expected benefit.

## 8.1 Plan Quality Management — Tools & Techniques 2

**Cost of quality.**
- **Prevention costs.** Costs related to the prevention of poor quality in the products, deliverables.
- **Appraisal costs.** Costs related to evaluating, measuring, auditing, and testing the products, deliverables.
- **Failure costs (internal/external).** Costs related to nonconformance of the products, deliverables, or services to the needs or expectations of the stakeholders.

The optimal COQ is one that reflects the appropriate balance for investing in the cost of prevention and appraisal to avoid failure costs.

| Cost of Conformance | | Cost of Nonconformance |
|---|---|---|
| **Prevention Costs** (Build a quality product)<br>• Training<br>• Document processes<br>• Equipment<br>• Time to do it right | **Internal Failure Costs** (Failures found by the project)<br>• Rework<br>• Scrap | |
| **Appraisal Costs** (Assess the quality)<br>• Testing<br>• Destructive testing loss<br>• Inspections | **External Failure Costs** (Failures found by the Customer)<br>• Liabilities<br>• Warranty work<br>• Lost business | |
| Money spent during the project **to avoid failures** | | Money spent during and after the project **because of failures** |

## 8.1 Plan Quality Management — Tools & Techniques 3

**04 DECISION MAKING**

**05 DATA REPRESENTATION**
Flowcharts.

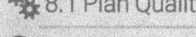

- Or process maps because they display the sequence of steps and the branching possibilities that exist for a process that transforms one or more inputs into one or more outputs.

- Flowcharts show the activities, decision points, branching loops, parallel paths, and the overall order of processing by mapping the operational details of procedures that exist within a horizontal value chain.

- SIPOC (suppliers, inputs, process, outputs, and customers) value chain model.

- Flowcharts may prove useful in understanding and estimating the cost of quality for a process.

- Process flows or process flow diagrams When flowcharts used to represent steps in a process, and they can be used for process improvement as well as identifying where quality defects can occur or where to incorporate quality checks.

## 8.1 Plan Quality Management  Tools & Techniques 4

- **Logical data model.** Visual representation of an organization's data, described in business language and independent of any specific technology.
The logical data model can be used to identify where data integrity or other quality issues can arise.

- **Matrix diagrams.** Matrix diagrams help find the strength of relationships among different factors, causes, and objectives that exist between the rows and columns that form the matrix.
Depending on how many factors may be compared, the project manager can use different shapes of matrix diagrams; for example, L, T, Y, X, C, and roof–shaped. In this process they facilitate identifying the key quality metrics that are important for the success of the project.

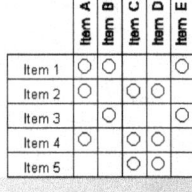

- **Mind mapping** The mind-mapping technique may help in the rapid gathering of project quality requirements, constraints, dependencies, and relationships.

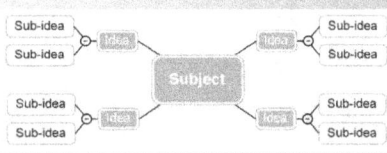

## 8.1 Plan Quality Management — Tools & Techniques 5

**06 TEST AND INSPECTION PLANNING**
PM and the project team determine how to test or inspect the product, deliverable, or service to meet the stakeholders' needs and expectations, as well as how to meet the goal for the product's performance and reliability.

**07 MEETINGS**

## 8.1 Plan Quality Management — Output 1

### QUALITY MANAGEMENT PLAN
- Describes how applicable policies, procedures, and guidelines will be implemented to achieve the quality objectives.
- It describes the activities and resources necessary for the project management team to achieve the quality objectives set for the project.
- It may be formal or informal, detailed, or broadly framed.
- It should be reviewed early in the project to ensure that decisions are based on accurate information.
- Provide Sharper focus on the project's value proposition, reductions in costs, and less frequent schedule overruns that are caused by rework.

**Includes:**
- Quality standards that will be used by the project;
- Quality objectives of the project;
- Quality roles and responsibilities;
- Project deliverables and processes subject to quality review;
- Quality control and quality management activities planned for the project;
- Quality tools that will be used for the project;
- Major procedures relevant for the project, such as dealing with nonconformance, corrective actions procedures, and continuous improvement procedures

## 8.1 Plan Quality Management — Output 2

### QUALITY METRICS
Specifically describes a project or product attribute and how the Control Quality process will verify compliance to it.

Include (percentage of tasks completed on time, cost performance measured by CPI, failure rate, number of defects identified per day, total downtime per month, errors found per line of code, customer satisfaction scores, and percentage of requirements covered by the test plan as a measure of test coverage.

### PROJECT MANAGEMENT PLAN UPDATES
- Risk management plan.
- Scope baseline

### PROJECT DOCUMENTS UPDATES
- Lessons learned register
- Requirements traceability matrix.
- Risk register.
- Stakeholder register.

## 8.2 Manage Quality

**MANAGE QUALITY**
Is the process of translating the quality management plan into executable quality activities that incorporate the organization's quality policies into the project.

**THE KEY BENEFIT**
It increases the probability of meeting the quality objectives as well as identifying ineffective processes and causes of poor quality.

- Manage Quality uses the data and results from the control quality process to reflect the overall quality status of the project to the stakeholders.
- This process is performed throughout the project

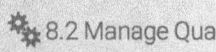

## 8.2 Manage Quality

- Sometimes called quality assurance,
- It is refeed as non-project work.
- QA focus on the processes used in the project.
- QA is about using project processes effectively.
- It involves following and meeting standards to assure the final product will meet stakeholders needs, expectations, and requirements.
- Includes all the quality assurance activities, and concerned with the product design aspects and process improvements.
- Fall under the conformance work category in the cost of quality framework.

helps to:
- Design an optimal and mature product by implementing specific design guidelines that address specific aspects of the product,
- Build confidence that a future output will be completed in a manner that meets the specified requirements and expectations through quality assurance tools and techniques such as quality audits and failure analysis,
- Confirm that the quality processes are used and that their use meets the quality objectives of the project, and
- Improve the efficiency and effectiveness of processes and activities to achieve better results and performance and enhance stakeholders' satisfaction.

## 8.2 Manage Quality

- The project manager and project team may use the organization's quality assurance department, to execute some of the Manage Quality activities such as failure analysis, design of experiments, and quality improvement.

- Quality assurance departments usually have cross-organizational experience in using quality tools and techniques and are a good resource for the project.

- Manage Quality is considered the work of everybody ( PM, project team, sponsor, management, and even the customer). Each has its roles in managing quality

- In agile projects, quality management is performed by all team members throughout the project,
  but in traditional projects, quality management is often the responsibility of specific team members.

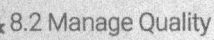

## 8.2 Manage Quality

### Inputs

.1 Project management plan
  • Quality management plan
.2 Project documents
  • Lessons learned register
  • Quality control measurements
  • Quality metrics
  • Risk report
.3 OPA

### Tools & Techniques

.1 Data gathering
  • Checklists
.2 Data analysis
  • Alternatives analysis
  • Document analysis
  • Process analysis
  • Root cause analysis
.3 Decision making
  • Multi-criteria decision analysis
.4 Data representation
  • Affinity diagrams
  • Cause-and-effect diagrams
  • Flowcharts
  • Histograms
  • Matrix diagrams
  • Scatter diagrams
.5 Audits
.6 Design for X
.7 Problem solving
.8 Quality improvement methods

### Outputs

.1 Quality reports
.2 Test and evaluation documents
.3 Change requests
.4 Project management plan updates
  • Quality management plan
  • Scope baseline
  • Schedule baseline
  • Cost baseline
.5 Project documents updates
  • Issue log
  • Lessons learned register
  • Risk register

## 8.2 Manage Quality  Input

**01 PROJECT MANAGEMENT PLAN**
- Quality management plan

**02 PROJECT DOCUMENTS**
- Lessons learned register
- Quality control measurements
- Quality metrics
- Risk report

**03 ORGANIZATIONAL PROCESS ASSETS**

## 8.2 Manage Quality — Tools & Techniques 1

### DATA GATHERING
**Checklist**
Is a structured tool used to verify that a set of required steps has been performed or to check if a list of requirements has been satisfied.
Checklists may be simple or complex.
Quality checklists should incorporate the acceptance criteria included in the scope baseline.

### DATA ANALYSIS
- Alternatives analysis. Select which different quality options or approaches are most appropriate to use.
- Document analysis. Such as quality reports, test reports, performance reports, and variance analysis.
- Process analysis. Identifies opportunities for process improvements.
- Root cause analysis **(RCA).** To determine the basic reason that causes a variance, defect, or risk, and solving them.

## 8.2 Manage Quality  Tools & Techniques 2

**DECISION MAKING**
Product decisions can include evaluating the life cycle cost, schedule, stakeholder satisfaction, and risks associated with resolving product defects.

**DATA REPRESENTATION**
- Affinity diagrams. Organize potential causes of defects into groups showing areas that should be focused on the most.
- Cause-and-effect diagrams. Or fishbone diagrams, why-why diagrams, or Ishikawa diagrams. This type of diagram breaks down the causes of the problem statement identified into discrete branches, helping to identify the main or root cause of the problem.
- Flowcharts. Show a series of steps that lead to a defect.
- Histograms. Graphical representation of numerical data. Histograms can show the number of defects per deliverable, a ranking of the cause of defects, the number of times each process is noncompliant,
- Matrix diagrams. Show the strength of relationships among factors, causes, and objectives that exist between the rows and columns that form the matrix.
- Scatter diagrams. Graph that shows the relationship between two variables.

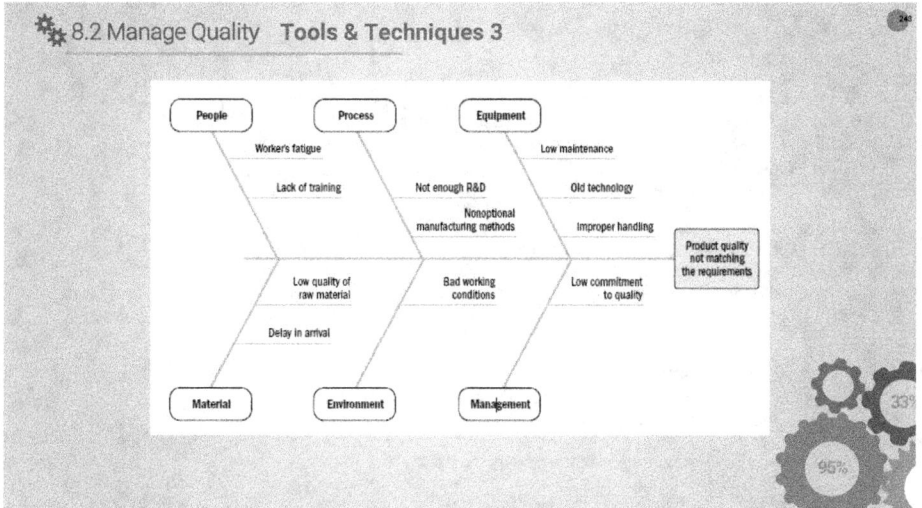

## 8.2 Manage Quality  Tools & Techniques 4

### AUDITS
An audit is a structured, independent process used to determine if project activities comply with organizational and project policies, processes, and procedures.
Usually conducted by a team external to the project (organization's internal audit department, PMO, or by an auditor external to the organization).
Quality audit objectives may include :
- Identifying all good and best practices being implemented;
- Identifying all nonconformity, gaps, and shortcomings;
- Sharing good practices introduced or implemented in similar projects in the Organization;
- Proactively offering assistance in a positive manner to improve the implementation of processes to help raise team productivity;
- Highlighting contributions of each audit in the lessons learned repository of the organization.

### QUALITY IMPROVEMENT METHODS
Quality improvements can occur based on findings and recommendations from quality control processes, the findings of the quality audits, or problem solving in the Manage Quality process.
Plan-do-check-act and Six Sigma are two of the most common quality improvement tools

## 8.2 Manage Quality  Tools & Techniques 5

### DESIGN FOR X
- Is a set of technical guidelines that may be applied during the design of a product for the optimization of a specific aspect of the design.
- DfX can control or even improve the product's final characteristics.
- The X in DfX can be different aspects of product development, such as reliability, deployment, assembly, manufacturing, cost, service, usability, safety, and quality.
- Using the DfX may result in cost reduction, quality improvement, better performance, and
- customer satisfaction.

### PROBLEM SOLVING
Problems can arise from Control Quality process or from quality audits and can be associated with a process or deliverable.
Problem-solving method will help eliminate the problem and develop a long-lasting solution. include:
1. Defining the problem,
2. Identifying the root-cause,
3. Generating possible solutions,
4. Choosing the best solution,
5. Implementing the solution, and
6. Verifying solution effectiveness.

## 8.2 Manage Quality — Output 1

**01 QUALITY REPORTS**
The quality reports can be graphical, numerical, or qualitative, include all quality management issues escalated by the team; recommendations for process, project, and product improvements; corrective actions recommendations (including rework, defect/bugs repair).

Can be used by other processes and departments to take corrective actions in order to achieve the project quality expectations.

**02 TEST AND EVALUATION DOCUMENTS**
They are inputs to the Control Quality process and are used to evaluate the achievement of quality objectives. may include dedicated checklists and detailed requirements traceability matrices as part of the document.

**03 CHANGE REQUESTS**

## 8.2 Manage Quality — Output 2

**Project management plan updates**
- Quality management plan
- Scope baseline
- Schedule baseline
- Cost baseline

**Project documents updates**
- Issue log
- Lessons learned register
- Risk register

## 8.3 Control Quality

**CONTROL QUALITY**
Is the process of monitoring and recording results of executing the quality management activities in order to assess performance and ensure the project outputs are complete, correct, and meet customer expectations.

**THE KEY BENEFIT**
verifying that project deliverables and work meet the requirements specified by key stakeholders for final acceptance

- Control Quality process is performed to measure the completeness, compliance, and fitness for use of a product or service prior to user acceptance and final delivery.

- In agile, Control Quality activities performed by all team members throughout the project life cycle.

- In waterfall model-based projects, quality control activities are performed at specific times, toward the end of the project or phase, by specified team members

## 8.3 Control Quality

### Inputs

.1 Project management plan
  • Quality management plan
.2 Project documents
  • Lessons learned register
  • Quality metrics
  • Test and evaluation documents
.3 Approved change requests
.4 Deliverables
.5 Work performance data
.6 EEF
.7 OPA

### Inputs Tools & Techniques Outputs

.1 Data gathering
  • Checklists
  • Check sheets
  • Statistical sampling
  • Questionnaires and surveys
.2 Data analysis
  • Performance reviews
  • Root cause analysis
.3 Inspection
.4 Testing/product evaluations
.5 Data representation
  • Cause-and-effect diagrams
  • Control charts
  • Histogram
  • Scatter diagrams
.6 Meetings

### Outputs

.1 Quality control measurements
.2 Verified deliverables
.3 Work performance information
.4 Change requests
.5 Project management plan updates
  • Quality management plan
.6 Project documents updates
  • Issue log
  • Lessons learned register
  • Risk register
  • Test and evaluation documents

## 8.3 Control Quality — Input

**01 Project management plan**
- Quality management plan

**02 Project documents**
- Lessons learned register
- Quality metrics
- Test and evaluation documents

**03 Approved change requests**

**04 Deliverables**

**05 Work performance data**

**06 Enterprise environmental factors**

**07 Organizational process assets**

## 8.3 Control Quality — Tools & Techniques 1

**01 DATA GATHERING**
- **Checklist**
- **Check sheets.** Or tally sheets, used to organize facts in a manner that will facilitate the effective collection of useful data about a potential quality problem.
- **Statistical sampling.** Involves choosing part of a population of interest for inspection (e.g. selecting 10 engineering drawings at random from a list of 75). to measure controls and verify quality.
- **Questionnaires and Surveys.** Gather data about customer satisfaction after the deployment of the product or service.

**02 DATA ANALYSIS**
- **Performance reviews.** Performance reviews measure, compare, and analyze the quality metrics defined by thePlan Quality Management process against the actual results.
- **Root cause analysis (RCA)**

**03 INSPECTION**
Is the examination of a work product to determine if it conforms to documented standards. Inspections may be called reviews, peer reviews, audits, or walkthroughs.

## 8.3 Control Quality — Tools & Techniques 2

**TESTING/PRODUCT EVALUATIONS**

Testing is an organized and constructed investigation conducted to provide objective information about the quality of the product or service in accordance with requirements.

- The intent of testing is to find errors, defects, bugs, or other nonconformance problems in the product or service.
- The type, amount, and extent of tests needed to evaluate each requirement are part of the project quality plan and depend on the nature of the project, time, budget, and other constraints.

- Tests can be performed throughout the project, as different components of the project become available, and at the end of the project on the final deliverables.

- Early testing helps identify nonconformance problems and helps reduce the cost of fixing the nonconforming components.

- Different application areas require different tests. For example, software testing may include unit testing,

## 8.3 Control Quality — Tools & Techniques 3

### DATA REPRESENTATION
- **Cause-and-effect diagrams**
- **Histograms**
- **Scatter diagrams.**
- **Control charts.**

Used to determine whether or not a process is stable or has predictable performance.
- Upper and lower specification limits are based on the requirements and reflect the maximum and minimum values allowed.
- Upper and lower control limits are different from specification limits.
- The control limits are determined using standard statistical calculations and principles to ultimately establish the natural capability for a stable process.
- The PM and SH use the statistically calculated control limits to identify the points at which corrective action will be taken to prevent performance that remains outside the control limits.

## 8.3 Control Quality  Tools & Techniques 3

**Control charts**
- Can be used to monitor various types of output variables.
- Used to track repetitive activities required for producing manufactured lots,
- Control charts may also be used to monitor cost and schedule variances, volume, frequency of scope changes, or other management results
- to help determine if the project management processes are in control.

### MEETINGS
**Approved change requests review.** All approved change requests should be reviewed to verify that they were implemented as approved.

**Retrospectives/lesson learned.** A meeting held by a project team to discuss:
- Successful elements in the project/phase,
- What could be improved,
- What to incorporate in the ongoing project and what in future projects, and
- What to add to the organization process assets.

## 8.3 Control Quality — Output 1

**01 QUALITY CONTROL MEASUREMENTS**
Are the documented results of Control Quality activities. They should be captured in the format that was specified in the quality management plan.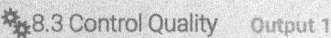

**02 VERIFIED DELIVERABLES**
The results of performing the Control Quality process are verified deliverables that become an input to the Validate Scope process for formalized acceptance.

**03 WORK PERFORMANCE INFORMATION**
Includes information on project requirements fulfillment, causes for rejections, rework required, recommendations for corrective actions, lists of verified deliverables, status of the quality metrics, and the need for process adjustments.

**04 CHANGE REQUESTS**

**05 PROJECT MANAGEMENT PLAN UPDATES**

**06 PROJECT DOCUMENTS UPDATES**
- Issue log.
- Lessons learned register.
- Risk register
- Test and evaluation documents.

# 9. PROJECT
## RESOURCE MANAGEMENT

Presented by:
**Nasser Al Mohimeed**
PMO Director, ISO 21500 Lead Project Manager
Certified Project Managers Trainer

## Project Resource Management

- Project Resource Management includes the processes to identify, acquire, and manage the resources needed for the successful completion of the project.

- Project Resource Management help ensure that the right resources will be available to the project manager and project team at the right time and place

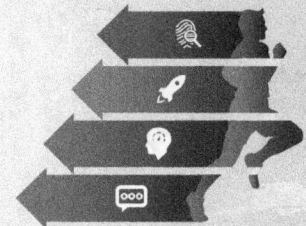

# Project Resource Management

| Initiating | Planning | Executing | Monitoring & Controlling | Closing |
|---|---|---|---|---|
| | 9.1 Plan Resource Management | 9.3 Acquire Resources | 9.6 Control Resources | |
| | 9.2 Estimate Activity Resources | 9.4 Develop Team | | |
| | | 9.5 Manage Team | | |

## Key concepts for Project Resource Management

- The project team consists of individuals with assigned roles and responsibilities who work collectively to achieve a shared project goal.
- PM should invest suitable effort in acquiring, managing, motivating, and empowering the project team.
- the involvement of all team members in project planning and decision making is beneficial. Participation of team members during planning adds their expertise to the process and strengthens their commitment to the project.
- PM should be both leader and manager of the project team. In addition to project management activities
- PM is responsible for the team formation as an effective group.
- PM also responsible for proactively developing team skills and competencies while retaining and improving team satisfaction and motivation.
- PM should be aware of, and subscribe to, professional and ethical behavior, and ensure that all team members adhere to these behaviors.

## Key concepts for Project Resource Management

- PM should be aware of different aspects that influence the team, such as:
  - Team environment,
  - Geographical locations of team members,
  - Communications among stakeholders,
  - Organizational change management,
  - Internal and external politics,
  - Cultural issues and organizational uniqueness, and
  - Other factors that may alter project performance.
- Physical resource management is concentrated in allocating and using the physical resources (material, equipment, and supplies, for example) needed for successful completion of the project in an efficient and effective way.
- Failing to manage and control resources efficiently is a source of risk for successful project completion.
- low inventory level, on the other hand, may result in not satisfying customer demand and, again, reduce the organization's profit.

## TRENDS AND EMERGING PRACTICES IN PROJECT RESOURCE MANAGEMENT

- Project management styles are shifting from a command and control structure for managing projects and toward a more collaborative and supportive management approach that empowers teams by delegating decision making to the team members.
Modern project resource management approaches seek to optimize resource utilization.

- **Resource management methods.** Due to the scarce nature of critical resources, in some industries, several trends have become popular
  - justin-time (JIT) manufacturing, Kaizen, total productive maintenance (TPM), theory of constraints (TOC), and other methods.
  - PM should determine and adapt the project accordingly.

- **Emotional intelligence (EI).** The project manager should invest in personal EI by improving inbound (e.g.,self-management and self-awareness) and outbound (e.g., relationship management) competencies.

- **Self-organizing teams.** using agile approaches ; rise to the self-organizing team, where the team functions with an absence of centralized control.
  - Consist of generalized specialists, instead of subject matter experts
  - PM role provides the team with the environment and support needed and trusts the team to get the job done.

## TRENDS AND EMERGING PRACTICES IN PROJECT RESOURCE MANAGEMENT

- **Virtual teams/distributed teams.** The globalization of projects has promoted the need for virtual teams that work on the same project, but are not co-located at the same site.

  - The availability of communication technology such as email, audio conferencing, social media, web-based meetings, and video conferencing has made virtual teams feasible.
  - Advantages,
    - being able to use special expertise on a project team even when the expert is not in the same geographic area,
    - incorporating employees who work from home offices,
    - including people with mobility limitations or disabilities.

  - The challenges of managing virtual teams are
    - mainly in the communication domain, including a possible feeling of isolation,
    - gaps in sharing knowledge and experience between team members,
    - and difficulties in tracking progress and productivity,
    - possible time zone difference
    - cultural differences.

# TAILORING CONSIDERATIONS

Each project is unique; therefore, the project manager will need to tailor the way Project Resources Management processes are applied

- **Diversity.** What is the diversity background of the team?
- **Physical location.** What is the physical location of team members and physical resources?
- **Industry-specific resources.** What special resources are needed in the industry?
- **Acquisition of team members.** How will team members be acquired for the project? Are team resources full-time or part-time on the project?
- **Management of team.** How is team development managed for the project? Are there organizational tools to manage team development or will new ones need to be established? Are there team members who have special needs? Will the team need special training to manage diversity?
- **Life cycle approaches.** What life cycle approach will be used on the project?

## CONSIDERATIONS FOR AGILE/ADAPTIVE ENVIRONMENTS

- Agile project benefit from team structures that maximize focus and collaboration, such as self organizing teams with generalizing specialists.
- Collaboration is intended to boost productivity and facilitate innovative problem solving.
- Collaborative teams may facilitate accelerated integration of distinct work activities, improve communication, increase knowledge sharing, and provide flexibility of work assignments in addition to other advantages.
- collaborative teams are often critical to the success of projects with a high degree of variability and rapid changes, because there is less time for centralized tasking and decision making.
- Planning for physical and human resources is much less predictable in projects with high variability.
- In these environments, agreements for fast supply and lean methods are critical to controlling costs and achieving the schedule.

## 9.1 Plan Resource Management

**PLAN RESOURCE MANAGEMENT** is the process of defining how to estimate, acquire, manage, and use team and physical resources.

**THE KEY BENEFIT** It establishes the approach and level of management effort needed for managing project resources based on the type and complexity of the project

 Resource planning is used to determine and identify an approach to ensure that sufficient resources are available for the successful completion of the project.

 Project resources may include team members, supplies, materials, equipment, services and facilities.

## 9.1 Plan Resource Management

| Inputs | Tools & Techniques | Outputs |
|---|---|---|
| .1 Project charter<br>.2 Project management plan<br>• Quality management plan<br>• Scope baseline<br>.3 Project documents<br>• Project schedule<br>• Requirements documentation<br>• Risk register<br>• Stakeholder register<br>.4 EEF<br>.5 OPA | .1 Expert judgment<br>.2 Data representation<br>• Hierarchical charts<br>• Responsibility assignment matrix<br>• Text-oriented formats<br>.3 Organizational theory<br>.4 Meetings | .1 Resource management plan<br>.2 Team charter<br>.3 Project documents updates<br>• Assumption log<br>• Risk register |

## 9.1 Plan Resource Management — Input

**01 Project charter**

**02 Project management plan**
- Quality management plan
- Scope baseline

**03 Project documents**
- Project schedule
- Requirements documentation
- Risk register
- Stakeholder register

**04 Enterprise environmental factors**

**05 Organizational process assets**

## 9.1 Plan Resource Management — Tools & Techniques

**01 EXPERT JUDGMENT**
- Negotiating for the best resources within the organization;
- Talent management and personnel development;
- Determining the preliminary effort level needed to meet project objectives;
- Determining reporting requirements based on the organizational culture;
- Estimating lead times required for acquisition, based on lessons learned and market conditions;
- Identifying risks associated with resource acquisition, retention, and release plans;
- Complying with applicable government and union regulations; and
- Managing sellers and the logistics effort to ensure materials and supplies are available when needed.

**02 ORGANIZATIONAL THEORY**
provides information regarding the way in which people, teams, and organizational units behave.
Effective use of common techniques identified in organizational theory can shorten the amount of time, cost, and effort needed to create the Plan Resource Management
It is important to recognize the organization's structure and culture impacts the project organizational structure.

**03 MEETINGS**

## 9.1 Plan Resource Management — Tools & Techniques

**DATA REPRESENTATION**
- ❖ Hierarchical charts. The traditional organizational chart structure can be used to show positions and relationships in a graphical, top-down format.
- WBS.
- Organizational breakdown structure (OBS). Is arranged according to an organization's existing departments, units, or teams, with the project activities or work packages listed under each department.
- Resource breakdown structure. Hierarchical list of team and physical resources related by category and resource type that is used for planning, managing and controlling project work.

- ❖ Assignment Matrix. A RAM shows the project resources assigned to each work package. It is used to illustrate the connections between work packages, or activities, and project team members.
- The matrix format shows all activities associated with one person and all people associated with one activity.
- This also ensures that there is only one person accountable for any one task to avoid confusion about who is ultimately in charge or has authority for the work.
- One example of a RAM is a RACI (responsible, accountable, consult, and inform)

## 9.1 Plan Resource Management — Tools & Techniques

| RACI Chart | Person | | | | |
|---|---|---|---|---|---|
| Activity | Ann | Ben | Carlos | Dina | Ed |
| Create charter | A | R | I | I | I |
| Collect requirements | I | A | R | C | C |
| Submit change request | I | A | R | R | C |
| Develop test plan | A | C | I | I | R |

R = Responsible  A = Accountable  C = Consult  I = Inform

- **Text-oriented formats.** Team member responsibilities that require detailed descriptions can be specified in text oriented formats.
- Usually in outline form, these documents provide information such as responsibilities, authority, competencies, and qualifications.
- The documents are known by various names including position descriptions and role-responsibility-authority forms.
- These documents can be used as templates for future projects, when the information is updated throughout project.

## 9.1 Plan Resource Management — Output 1

**RESOURCE MANAGEMENT PLAN**

provides guidance on how project resources should be categorized, allocated, managed, and released. management plan may include :

- ❖ **Identification of resources.** Methods for identifying and quantifying team and physical resources needed.
- ❖ **Acquiring resources.** Guidance on how to acquire team and physical resources for the project.
- ❖ **Roles and responsibilities**
  - Role. The function assumed by, or assigned to, a person in the project.
  - Authority. The rights to apply project resources, make decisions, sign approvals, accept deliverables,
  - Responsibility. The assigned duties and work that a project team member is expected to perform in order to complete the project's activities.
  - Competence. The skill and capacity required to complete assigned activities within the project constraints.
- ❖ **Project organization charts.** Is a graphic display of project team members and their reporting relationships. It can be formal or informal, highly detailed or broadly framed, based on the needs of the project.

## 9.1 Plan Resource Management  Output 2

- **Project team resource management.** Guidance on how project team resources should be defined, staffed, managed, and eventually released.
- **Training.** Training strategies for team members.
- **Team development.** Methods for developing the project team.
- **Resource control.** Methods for ensuring adequate physical resources are available as needed and that the acquisition of physical resources is optimized for project needs. Includes information on managing inventory, equipment, and supplies during throughout the project life cycle.
- **Recognition plan.** Which recognition and rewards will be given to team members, and when they will be given.

### TEAM CHARTER
is a document that establishes the team values, agreements, and operating guidelines for the team.
- Team values,
- Communication guidelines,
- Decision-making criteria and process,
- Conflict resolution process,
- Meeting guidelines, and
- Team agreements.

## 9.1 Plan Resource Management — Output 3

**PROJECT DOCUMENTS UPDATES**
- Assumption log
- Risk register

## 9.2 Estimate Activity Resources

**ESTIMATE ACTIVITY RESOURCES**
is the process of estimating team resources and the type and quantities of materials, equipment, and supplies necessary to perform project work.

**THE KEY BENEFIT**
is that it identifies the type, quantity, and characteristics of resources required to complete the project.

This process is performed periodically throughout the project as needed

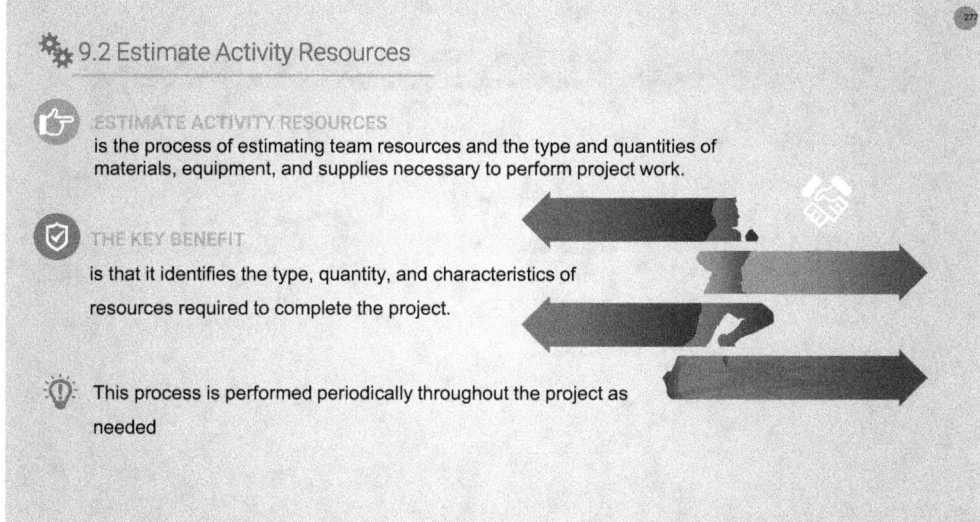

## 9.2 Estimate Activity Resources

### Inputs

.1 Project management plan
  • Resource management plan
  • Scope baseline
.2 Project documents
  • Activity attributes
  • Activity list
  • Assumption log
  • Cost estimates
  • Resource calendars
  • Risk register
.3 EEF
.4 OPA

### Tools & Techniques

.1 Expert judgment
.2 Bottom-up estimating
.3 Analogous estimating
.4 Parametric estimating
.5 Data analysis
  • Alternatives analysis
.6 PMIS
.7 Meetings

### Outputs

.1 Resource requirements
.2 Basis of estimates
.3 Resource breakdown structure
.4 Project documents updates
  • Activity attributes
  • Assumption log
  • Lessons learned register

## 9.2 Estimate Activity Resources — Input

**01 PROJECT MANAGEMENT PLAN**
- Resource management plan
- Scope baseline

**02 PROJECT DOCUMENTS**
- Activity attributes
- Activity list
- Assumption log
- Cost estimates
- Resource calendars
- Risk register

**03 Enterprise environmental factors**

**04 Organizational process assets**

## 9.2 Estimate Activity Resources — Tools & Techniques

01. **EXPERT JUDGMENT**
02. **BOTTOM-UP ESTIMATING**
03. **ANALOGOUS ESTIMATING**
04. **PARAMETRIC ESTIMATING**
05. **DATA ANALYSIS**
    - Alternatives analysis
06. **PROJECT MANAGEMENT INFORMATION SYSTEM**
07. **MEETINGS**

## 9.2 Estimate Activity Resources — Output 1

**RESOURCE REQUIREMENTS**
Identify the types and quantities of resources required for each work package or activity in a work package and can be aggregated to determine the estimated resources for each work package, each WBS branch, and the project as a whole.

**BASIS OF ESTIMATES**
The amount and type of additional details supporting the resource estimate vary by application area.
- Method used to develop the estimate,
- Resources used to develop the estimate (such as information from previous similar projects),
- Assumptions associated with the estimate,
- Known constraints,
- Range of estimates,
- Confidence level of the estimate, and
- Documentation of identified risks influencing the estimate.

## 9.2 Estimate Activity Resources — Output 2

**RESOURCE BREAKDOWN STRUCTURE**
is a hierarchical representation of resources by category and type
Categories include but are not limited to labor, material, equipment, and supplies.
Type include the skill level, grade level, required certifications, or other information.

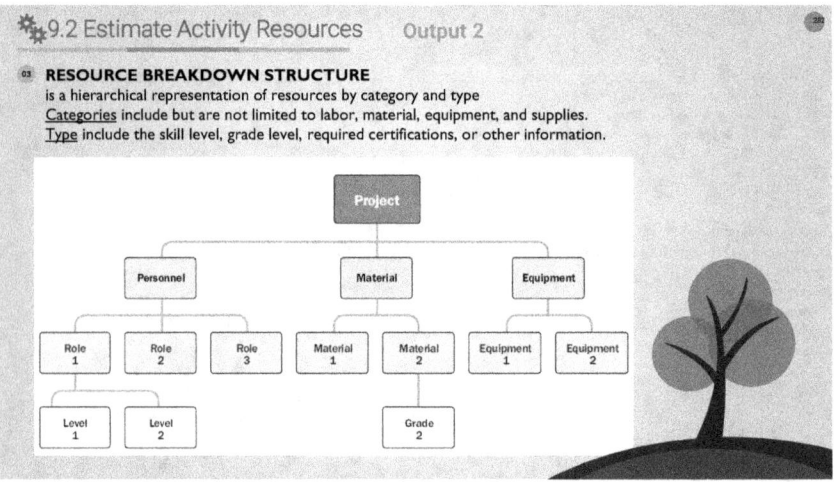

## 9.2 Estimate Activity Resources — Output 3

**PROJECT DOCUMENTS UPDATES**
- Activity attributes
- Assumption log
- Lessons learned register

## 9.3 Acquire Resources

**ACQUIRE RESOURCES**
Is the process of obtaining team members, facilities, equipment, materials, supplies, and other resources necessary to complete project work.

**THE KEY BENEFIT**
It outlines and guides the selection of resources and assigns them to their respective activities.

 This process is performed periodically throughout the project as needed

## 9.3 Acquire Resources

The resources needed for the project can be internal or external to the project-performing organization.
- Internal resources are acquired (assigned) from functional or resource managers.
- External resources are acquired through the procurement processes.

important factors to be considered during acquiring resources:
- The project manager or project team should effectively negotiate and influence others who are in a position to provide the required team and physical resources for the project.
- Failure to acquire the necessary resources for the project may affect project schedules, budgets, customer satisfaction, quality, and risks. Insufficient resources or capabilities decrease the probability of success and, in a worst-case scenario, could result in project cancellation.
- If the team resources are not available due to constraints such as economic factors or assignment to other projects, the project manager or project team may be required to assign alternative resources. Alternative resources are allowed provided there is no violation of legal, regulatory, mandatory, or other specific criteria.

The project manager or project management team will be required to document the impact of the unavailability of required resources in the project schedule, project budget, project risks, project quality, training plans, and other project management plans

## 9.3 Acquire Resources

### Inputs

.1 Project management plan
  • Resource management plan
  • Procurement management plan
  • Cost baseline
.2 Project documents
  • Project schedule
  • Resource calendars
  • Resource requirements
  • Stakeholder register
.3 EEF
.4 OPA

### Inputs Tools & Techniques Outputs

.1 Decision making
  • Multicriteria decision analysis
.2 Interpersonal and team skills
  • Negotiation
.3 Pre-assignment
.4 Virtual teams

### Outputs

.1 Physical resource assignments
.2 Project team assignments
.3 Resource calendars
.4 Change requests
.5 Project management plan updates
  • Resource management plan
  • Cost baseline
.6 Project documents updates
  • Lessons learned register
  • Project schedule
  • Resource breakdown structure
  • Resource requirements
  • Risk register
  • Stakeholder register
.7 EEF updates
.8 OPA Updates

## 9.3 Acquire Resources — Input

**01 PROJECT MANAGEMENT PLAN**
- Resource management plan
- Procurement management plan
- Cost baseline

**02 PROJECT DOCUMENTS**
- Project schedule
- Resource calendars
- Resource requirements
- Stakeholder register

**03 Enterprise environmental factors**

**04 Organizational process assets**

## 9.3 Acquire Resources — Tools & Techniques 1

### DECISION MAKING
Using a multi-criteria decision analysis tool, criteria are developed and used to rate or score potential. The criteria are weighted according to their relative importance and values can be changed for different types of resources.

Examples of selection criteria that can be used are
- **Availability.** Verify that the resource is available to work on the project within the time period needed.
- **Cost.** Verify if the cost of adding the resource is within the prescribed budget.
- **Ability.** Verify that the team member provides the capability needed by the project.

Unique selection criteria that for team resources are:
- **Experience.** Verify that the team member has the relevant experience
- **Knowledge.** Consider if the team member has relevant knowledge of the customer, similar implemented projects, and nuances of the project environment.
- **Skills.** Determine if the team member has the relevant skills to use a project tool.
- **Attitude.** Determine if the team member has the ability to work with others as a cohesive team.
- **International factors.** Consider team member location, time zone, and communication capabilities.

## 9.3 Acquire Resources — Tools & Techniques 2

### INTERPERSONAL AND TEAM SKILLS
The project management team may need to negotiate with:
- **Functional managers.** Ensure that the project receives the best resources possible in the required timeframe and until their responsibilities are complete.
- **Other project management teams within the performing organization.** On scarce or specialized resources.
- **External organizations and suppliers.** Provide appropriate, scarce, specialized, qualified, certified, or other specific team or physical resources. Special consideration should be given to external negotiating policies, practices, processes, guidelines, legal, and other such criteria.

### PRE-ASSIGNMENT
When physical or team resources for a project are determined in advance, they are considered pre-assigned. (when resources being identified as part of a competitive proposal, or if the project is dependent upon the expertise of particular persons).

## 9.3 Acquire Resources — Tools & Techniques 3

### 04 VIRTUAL TEAMS

Virtual teams can be defined as groups of people with a shared goal who fulfill their roles with little or no time spent meeting face to face. The availability of communication technology such as email, audio conferencing, social media, web-based meetings, and video conferencing has made virtual teams feasible.

The virtual team model makes it possible to:
- Form teams of people from the same organization who live in widespread geographic areas;
- Add special expertise to a project team even though the expert is not in the same geographic area;
- Incorporate employees who work from home offices;
- Form teams of people who work different shifts, hours, or days;
- Include people with mobility limitations or disabilities;
- Move forward with projects that would have been held or canceled due to travel expenses; and
- Save the expense of offices and all physical equipment needed for employees.

## 9.3 Acquire Resources — Output 1

**01 PHYSICAL RESOURCE ASSIGNMENTS**
Documentation of the physical resource assignments records the material, equipment, supplies, locations, and other physical resources that will be used during the project.

**02 PROJECT TEAM ASSIGNMENTS**
Documentation of team assignments records the team members and their roles and responsibilities for the project.
Documentation can include a project team directory and names inserted into the project management plan, such as the project organization charts and schedules.

**03 RESOURCE CALENDARS**
Identifies the working days, shifts, start and end of normal business hours, weekends, and public holidays when each specific resource is available.
Also specify when and for how long identified team and physical resources will be available during the project.

**04 CHANGE REQUESTS**

## 9.3 Acquire Resources — Output 2

**PROJECT MANAGEMENT PLAN UPDATES**
- Resource management plan.
- Cost baseline.

**PROJECT DOCUMENTS UPDATES**
- Lessons learned register.
- Project schedule
- Resource breakdown structure.
- Resource requirements.
- Risk register.
- Stakeholder register.

**ENTERPRISE ENVIRONMENTAL FACTORS UPDATES**
- Resource availability within the organization, and
- Amount of the organization's consumable resources that have been used.

**ORGANIZATIONAL PROCESS ASSETS UPDATES**
include updates to documentation related to acquiring, assigning and allocating resources.

## 9.4 Develop Team

**DEVELOP TEAM**
is the process of improving competencies, team member interaction, and the overall team environment to enhance project performance.

**THE KEY BENEFIT**
it results in improved teamwork, enhanced interpersonal skills and competencies, motivated employees, reduced attrition, and improved overall project performance.

This process is performed periodically throughout the project as needed

## 9.4 Develop Team

- Project managers require the skills to identify, build, maintain, motivate, lead, and inspire project teams to achieve high team performance and to meet the project's objectives.

- Teamwork is a critical factor for project success, and

- PM should create an environment that facilitates teamwork and continually motivates the team by providing challenges and opportunities, providing timely feedback and support as needed, and recognizing and rewarding good performance.

- High team performance can be achieved by employing these behaviors:
  - Using open and effective communication,
  - Creating team-building opportunities,
  - Developing trust among team members,
  - Managing conflicts in a constructive manner,
  - Encouraging collaborative problem solving, and
  - Encouraging collaborative decision making.

- Developing the project requires clear, timely, effective, and efficient communication between team members throughout the life of the project.

## 9.4 Develop Team

Objectives of developing a project team :
- Improving the knowledge and skills of team members to increase their ability to complete project deliverables, while lowering costs, reducing schedules, and improving quality;
- Improving feelings of trust and agreement among team members to raise morale, lower conflict, and increase teamwork;
- Creating a dynamic, cohesive, and collaborative team culture to:
   (1) improve individual and team productivity, team spirit, and cooperation;
   (2) allow cross-training and mentoring between team members to share knowledge and expertise;

- Empowering the team to participate in decision making and take ownership of the provided solutions to improve team productivity for more effective and efficient results.

## 9.4 Develop Team

**Tuckman ladder model**

- **Forming.** Team members meet and learn about the project and their formal roles and responsibilities.

- **Storming.** Team begins to address the project work, technical decisions, and the project management approach

- **Norming.** Team members begin to work together and adjust their work habits and behaviors to support the team. The team members learn to trust each other.

- **Performing.** Teams that reach the performing stage function as a well-organized unit. They are interdependent and work through issues smoothly and effectively.

- **Adjourning.** The team completes the work and moves on from the project. This typically occurs when staff is released from the project as deliverables are completed or as part of the Close Project or Phase process

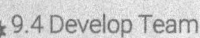

## 9.4 Develop Team

**Inputs**

.1 Project management plan
 • Resource management plan
.2 Project documents
 • Lessons learned register
 • Project schedule
 • Project team assignments
 • Resource calendars
 • Team charter
.3 EEF
.4 OPA

**Tools & Techniques**

.1 Colocation
.2 Virtual teams
.3 Communication technology
.4 Interpersonal and team skills
 • Conflict management
 • Influencing
 • Motivation
 • Negotiation
 • Team building
.5 Recognition and rewards
.6 Training
.7 Individual and team assessments
.8 Meetings

**Outputs**

1. Team performance assessments
.2 Change requests
.3 Project management plan updates
 • Resource management plan
.4 Project documents updates
 • Lessons learned register
 • Project schedule
 • Project team assignments
 • Resource calendars
 • Team charter
.5 EEF updates
.6 OPA Updates

## 9.4 Develop Team  Input

**01 PROJECT MANAGEMENT PLAN**
- Resource management plan

**02 PROJECT DOCUMENTS**
- Lessons learned register
- Project schedule
- Project team assignments
- Resource calendars
- Team charter

**03 Enterprise environmental factors**

**04 Organizational process assets**

## 9.4 Develop Team — Tools & Techniques 1

**01 COLOCATION**
involves placing many or all of the most active project team members in the same physical location to enhance their ability to perform as a team. include a team meeting room, common places to post schedules, and other conveniences that enhance communication and a sense of community.

**02 VIRTUAL TEAMS**

**03 COMMUNICATION TECHNOLOGY**
- Shared portal.
- conferencing.
- Audio conferencing.
- Email/chat..

**04 INTERPERSONAL AND TEAM SKILLS**
- **Conflict management**
- **Influencing.**
- **Motivation.** Motivation is providing a reason for someone to act
- **Negotiation.**
- **Team building.** Conducting activities that enhance the team's social relations and build a collaborative and cooperative working environment

## 9.4 Develop Team — Tools & Techniques 2

### RECOGNITION AND REWARDS
- Part of the team development process involves recognizing and rewarding desirable behavior.
- Rewards will be effective only if they satisfy a need that is valued by that individual.
- Reward decisions are made, formally or informally,

### TRAINING
includes all activities designed to enhance the competencies of the project team members.
Can be formal or informal.

### INDIVIDUAL AND TEAM ASSESSMENTS
give the project manager and the project team insight into areas of strengths and weaknesses.
help project managers assess team members' preferences, aspirations, how they make decisions,

### MEETINGS

## 9.4 Develop Team — Output 1

**01 TEAM PERFORMANCE ASSESSMENTS**
Effective team development strategies and activities are expected to increase the team's performance, which increases of meeting project objectives.
**The evaluation of a team's effectiveness may include indicators such as:**
- Improvements in skills that allow individuals to perform assignments more effectively,
- Improvements in competencies that help team members perform better as a team,
- Reduced staff turnover rate, and
- Increased team cohesiveness where team members share information and experiences openly and help each other to improve the overall project performance.

As a result of conducting an evaluation of the, the project management team can identify the specific training, coaching, mentoring, assistance, or changes required to improve the team's performance.

**02 CHANGE REQUESTS**

**03 PROJECT MANAGEMENT PLAN UPDATES**

## 9.4 Develop Team — Output 2

**PROJECT DOCUMENTS UPDATES**
- Lessons learned register
- Project schedule.
- Project team assignments
- Resource calendars
- Team charter.

**ENTERPRISE ENVIRONMENTAL FACTORS UPDATES**
- Employee development plan records
- Skill assessments.

**ORGANIZATIONAL PROCESS ASSETS UPDATES**
- Training requirements
- Personnel assessment.

## 9.5 MANAGE TEAM

**MANAGE TEAM**
is the process of tracking team member performance, providing feedback, resolving issues, and managing team changes to optimize project performance.

**THE KEY BENEFIT**
it influences team behavior, manages conflict, and resolves issues.

This process is performed throughout the project as needed

## 9.5 MANAGE TEAM

**Inputs**
.1 Project management plan
  • Resource management plan
.2 Project documents
  • Issue log
  • Lessons learned register
  • Project team assignments
  • Team charter
.3 Work performance reports
.4 Team performance Assessments
.5 EEF
.6 OPA

**Inputs Tools & Techniques Outputs**
.1 Interpersonal and team skills
  • Conflict management
  • Decision making
  • Emotional intelligence
  • Influencing
  • Leadership
.2 PMIS

**Outputs**
.1 Change requests
.2 Project management plan updates
  • Resource management plan
  • Schedule baseline
  • Cost baseline
.3 Project documents updates
  • Issue log
  • Lessons learned register
  • Project team assignments
.4 EEF updates

## 9.5 MANAGE TEAM  Input

**01 PROJECT MANAGEMENT PLAN**
- Resource management plan

**02 PROJECT DOCUMENTS**
- Issue log
- Lessons learned register
- Project team assignments
- Team charter

**03 WORK PERFORMANCE REPORTS**

**04 TEAM PERFORMANCE ASSESSMENTS**

**05** Enterprise environmental factors

**06** Organizational process assets

## 9.5 MANAGE TEAM    Tools & Techniques 1

### INTERPERSONAL AND TEAM SKILLS
**Conflict management.**
Sources of conflict include scarce resources, scheduling priorities, and personal work styles. Team ground rules, group norms, and solid project management practices,
- Successful conflict management results in greater productivity and positive working relationships.
- Project team members are initially responsible for their resolution.
- If conflict escalates, PM should help facilitate a satisfactory resolution.

Conflict resolution methods include:
- Importance and intensity of the conflict,
- Time pressure for resolving the conflict,
- Relative power of the people involved in the conflict,
- Importance of maintaining a good relationship, and
- Motivation to resolve conflict on a long-term or short-term basis.

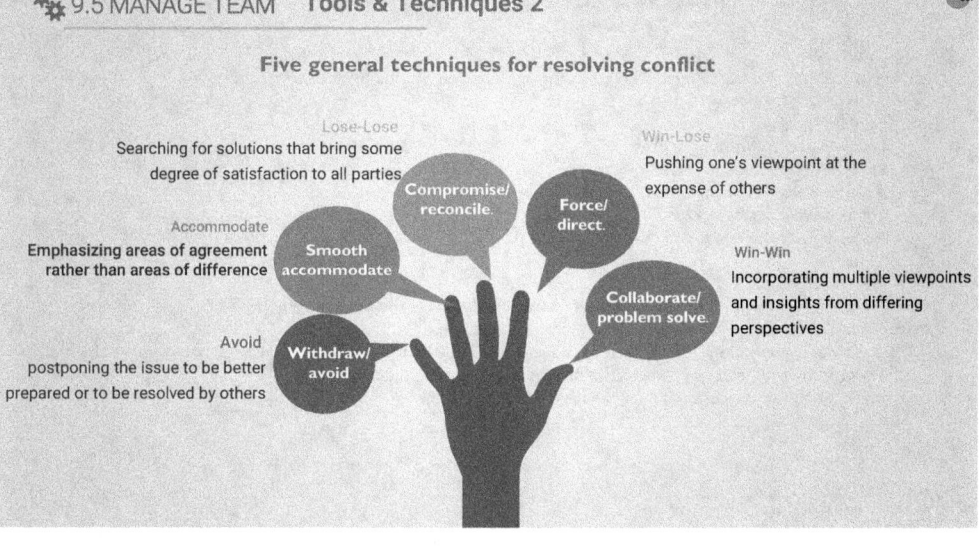

## 9.5 MANAGE TEAM — Tools & Techniques 3

**Decision making.**
involves the ability to negotiate and influence the organization and the project management team,
Decision making include:
- Focus on goals to be served,
- Follow a decision-making process,
- Study the environmental factors,
- Analyze available information,
- Stimulate team creativity, and
- Account for risk.

**Emotional intelligence.**
Emotional intelligence is the ability to identify, assess, and manage the personal emotions of oneself and other people, as well as the collective emotions of groups of people.
The team can use emotional intelligence to reduce tension and increase cooperation by identifying, assessing, and controlling the sentiments of project team members, anticipating their actions, acknowledging their concerns, and following up on their issues.

## 9.5 MANAGE TEAM — Tools & Techniques 4

**Influencing.**
Because project managers often have little or no direct authority over team members in a matrix environment, their ability to influence stakeholders on a timely basis is critical to project success. skills include:
- Ability to be persuasive;
- Clearly articulating points and positions;
- High levels of active and effective listening skills;
- Awareness of, and consideration for, the various perspectives in any situation; and
- Gathering relevant information to address issues and reach agreements while maintaining mutual trust.

**Leadership.** is the ability to lead a team and inspire them to do their jobs well.
Leadership is important through all phases of the project life cycle.

### PROJECT MANAGEMENT INFORMATION SYSTEM (PMIS)

## 9.5 MANAGE TEAM — Output

**01 CHANGE REQUESTS**

**02 PROJECT MANAGEMENT PLAN UPDATES**
- Resource management plan
- Schedule baseline
- Cost baseline

**03 PROJECT DOCUMENTS UPDATES**
- Issue log
- Lessons learned register
- Project team assignments

**04 ENTERPRISE ENVIRONMENTAL FACTORS UPDATES**

## 9.6 Control Resources

**CONTROL RESOURCES**
is the process of ensuring that the physical resources assigned and allocated to the project are available as planned, as well as monitoring the planned versus actual utilization of resources and taking corrective action as necessary.

**THE KEY BENEFIT**
is ensuring that the assigned resources are available to the project at the right time and in the right place and are released when no longer needed

This process is performed throughout the project as needed

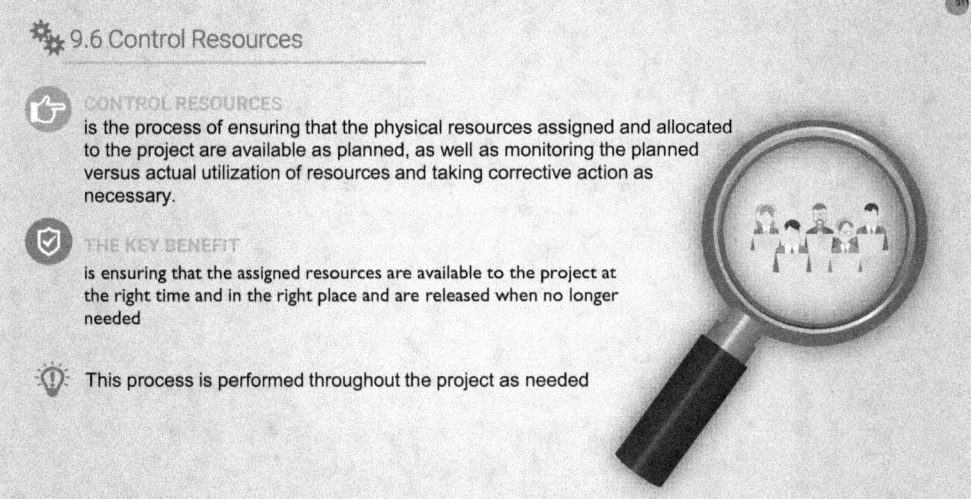

## 9.6 Control Resources

**Control Resources is concerned with:**
- Monitoring resource expenditures,
- Identifying and dealing with resource shortage/surplus in a timely manner,
- Ensuring that resources are used and released according to the plan and project needs,
- Informing appropriate stakeholders if any issues arise with relevant resources,
- Influencing the factors that can create resources utilization change, and
- Managing the actual changes as they occur.

## 9.6 Control Resources

### Inputs

.1 Project management plan
  • Resource management plan
.2 Project documents
  • Issue log
  • Lessons learned register
  • Physical resource assignments
  • Project schedule
  • Resource breakdown structure
  • Resource requirements
  • Risk register
.3 Work performance data
.4 Agreements
.5 OPA

### Tools & Techniques

.1 Data analysis
  • Alternatives analysis
  • Cost-benefit analysis
  • Performance reviews
  • Trend analysis
.2 Problem solving
.3 Interpersonal and team skills
  • Negotiation
  • Influencing
.4 PMIS

### Outputs

.1 Work performance information
.2 Change requests
.3 Project management plan updates
  • Resource management plan
  • Schedule baseline
  • Cost baseline
.4 Project documents updates
  • Assumption log
  • Issue log
  • Lessons learned register
  • Physical resource assignments
  • Resource breakdown structure
  • Risk register

## 9.6 Control Resources — Input

**01 PROJECT MANAGEMENT PLAN**
- Resource management plan

**02 PROJECT DOCUMENTS**
- Issue log
- Lessons learned register
- Physical resource assignments
- Project schedule
- Resource breakdown structure
- Resource requirements
- Risk register

**03 WORK PERFORMANCE DATA**

**04 AGREEMENTS**

**05 Organizational process assets**

## 9.6 Control Resources — Tools & Techniques 1

**01 DATA ANALYSIS**
- **Alternatives analysis**
- **Cost-benefit analysis**
- **Performance reviews.** compare, and analyze planned resource utilization to actual resource utilization.
- **Trend analysis.** examines project performance over time and can be used to determine whether performance is improving or deteriorating.

**02 PROBLEM SOLVING**
The project manager should use methodical steps to deal with problem solving,
- **Identify the problem.** Specify the problem.
- **Define the problem.** Break it into smaller, manageable problems.
- **Investigate.** Collect data.
- **Analyze.** Find the root cause of the problem.
- **Solve.** Choose the suitable solution from a variety of available ones.
- **Check the solution.** Determine if the problem has been fixed.

## 9.6 Control Resources — Tools & Techniques 2

**03 INTERPERSONAL AND TEAM SKILLS**
sometimes known as "soft skills," include:
- Negotiation.
- Influencing.

**04 PROJECT MANAGEMENT INFORMATION SYSTEM (PMIS)**

## 9.6 Control Resources — Output

**01 WORK PERFORMANCE INFORMATION**

**02 CHANGE REQUESTS**

**03 PROJECT MANAGEMENT PLAN UPDATES**
- Resource management plan
- Schedule baseline
- Cost baseline

**04 PROJECT DOCUMENTS UPDATES**
- Assumption log
- Issue log
- Lessons learned register
- Physical resource assignments
- Resource breakdown structure
- Risk register

Have You Any
# question ?

318

| What | Why | When | Where | How | who |

# 10. PROJECT
## COMMUNICATIONS MANAGEMENT

Presented by :
**Nasser Al Mohimeed**
PMO Director, ISO 21500 Lead Project Manager
Certified Project Managers Trainer

## Project Communications Management

**What is a Project Communications Management?**
Includes the processes necessary to ensure that the information needs of the project and its stakeholders are met through development of artifacts and implementation of activities designed to achieve effective information exchange

**Consists of two parts**
The first part is developing a strategy to ensure communication is effective for stakeholders.
The second part is carrying out the activities necessary to implement the communication strategy

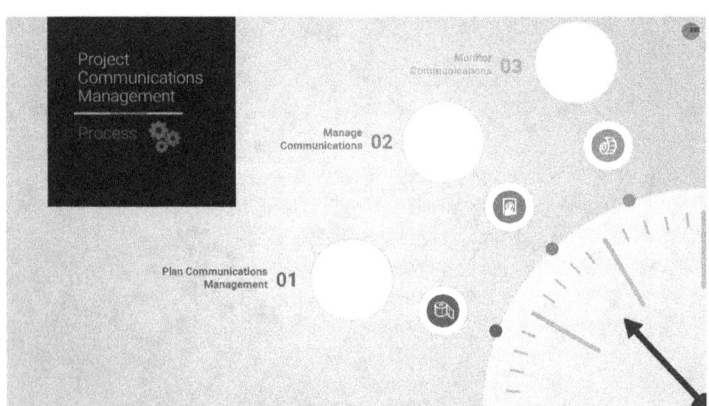

## Project Communications Management

| Initiating | Planning | Executing | Monitoring & Controlling | Closing |
|---|---|---|---|---|
| ⏻ | 📅 |  |  |  |
| | 10.1 Plan Communications Management | 10.2. Manage Communications | 10.3 Monitor Communications | |

## Key concepts for project communication management

- Project managers spend most of their time communicating with team members and other project stakeholders, both internal (at all organizational levels) and external to the organization

- Communication is the exchange of information, intended or involuntary. The information exchanged can be in the form of ideas, instructions, or emotions.

- Information can be sent or received, either through communication activities, such as meetings and presentations, or artifacts, such as emails, social media, project reports, or project documentation.

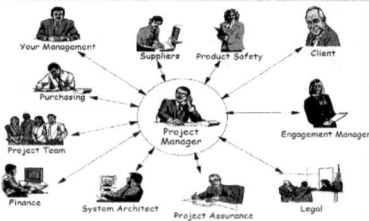

# Key concepts for project communication management

 **Mechanisms of exchange Information**

- **Written form.** Either physical or electronic.
- **Spoken.** Either face-to-face or remote.
- **Formal or informal** (as in formal papers or social media).
- **Through gestures.** Tone of voice and facial expressions.
- **Through media.** Pictures, actions, or even just the choice of words.
- **Choice of words.** There is often more than one word to express an idea;

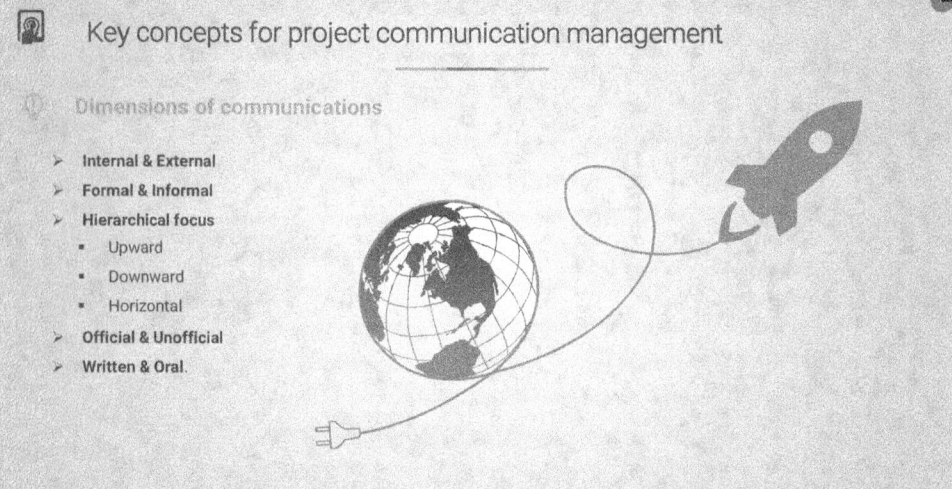

## 10.1 Plan communication management

**Is the process** of developing an appropriate approach and plan for project communications activities based on the information needs of each stakeholder or group, available organizational assets, and the needs of the project.

**The key benefit** of this process is a documented approach to effectively and efficiently engage stakeholders by presenting relevant information in a timely manner.

- Planning the project communications is important to the ultimate success of any project.
- Mostly performed during the early stages of the project, i.e., during developing project management plan.
- Is linked with enterprise environmental factors, because project's organizational structure impacts the project's communication requirements.
- The results of this process should be reviewed regularly throughout the project and revised as needed to ensure continued applicability.

## 10.1 Plan communication management

**Inputs**

.1 Project charter
.2 Project management plan
  • Resource management plan
  • Stakeholder engagement plan
.3 Project documents
  • Requirements documentation
  • Stakeholder register
.4 EEF
.5 OPA

**Inputs Tools & Techniques Outputs**

.1 Expert judgment
.2 Communication requirements analysis
.3 Communication technology
.4 Communication models
.5 Communication methods
.6 Interpersonal and team skills
  • Communication styles assessment
  • Political awareness
  • Cultural awareness
.7 Data representation
  • Stakeholder engagement assessment matrix
.8 Meetings

**Outputs**

.1 Communications management plan
.2 Project management plan updates
  • Stakeholder engagement plan
.3 Project documents updates
  • Project schedule
  • Stakeholder register

# 10.1 Plan communication management

**Input:**

1. Project charter
2. Project management plan
   - Resource management plan
   - Stakeholder engagement plan
3. Project documents
   - Requirements documentation
   - Stakeholder register
4. Enterprise environmental factors
5. Organizational process assets

## 10.1 Plan communication management

### Tools & Techniques (1/4)

**01** Expert judgment

**02** Communication requirements analysis
determines the information needs of the project stakeholders.

define project communication requirements:
- Number of potential communication channels
  ( one-to-one, one-to-many, and many-to-many)
- Organizational charts;
- Stakeholder responsibility, relationships, and interdependencies;
- Development approach;
- Disciplines, departments, and specialties involved in the project;
- Logistics of how many stockholders and their locations;
- Internal and External information
- Legal requirements.

## 10.1 Plan communication management

### Tools & Techniques (2/4)

**Communication technology**
The methods used to transfer information among stakeholders.

Factors affect the choice of communication technology
- Urgency of the need for information
- Availability and reliability of technology
- Ease of use.
- Project environment
- Sensitivity and confidentiality of the information.

# 10.1 Plan communication management

## Tools & Techniques (3/4)

**Communication models**

Sample basic sender/receiver communication model.

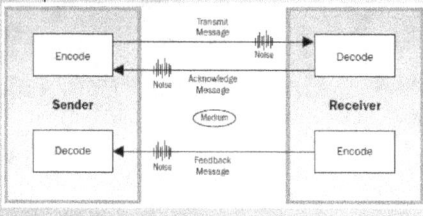

Semple interactive communication model

# 10.1 Plan communication management

## Tools & Techniques (4/4)

- **Communication Method**
  - Interactive communication
  - Push communication
  - Pull communication.

- **Interpersonal and team skills**
  - Communication styles assessment
  - Political awareness
  - Cultural awareness

- **Data representation**

- **Meetings**

# 10.1 Plan communication management

 **Output:**

**01 Communications management plan**

**02 Project management plan updates**
- Stakeholder engagement plan

**03 Project documents updates**
- Project schedule
- Stakeholder register

# 10.2 Manage Communications

**Is the process of** ensuring timely and appropriate collection, creation, distribution, storage, retrieval, management, monitoring, and the ultimate disposition of project information

**The key benefit** of this process is that it enables an efficient and effective information flow between the project team and the stakeholders

## 10.2 Manage Communications

| Inputs | Tools & Techniques | Outputs |
|---|---|---|
| .1 Project management plan<br>• Resource management plan<br>• Communications management plan<br>• Stakeholder engagement plan<br>.2 Project documents<br>• Change log<br>• Issue log<br>• Lessons learned register<br>• Quality report<br>• Risk report<br>• Stakeholder register<br>.3 Work performance reports<br>.4 EEF<br>.5 OPA | .1 Communication technology<br>.2 Communication methods<br>.3 Communication skills<br>• Communication competence<br>• Feedback<br>• Nonverbal<br>• Presentations<br>.4 PMIS<br>.5 Project reporting<br>.6 Interpersonal and team skills<br>• Active listening<br>• Conflict management<br>• Cultural awareness<br>• Meeting management<br>• Networking<br>• Political awareness<br>.7 Meetings | .1 Project communications<br>.2 Project management plan updates<br>• Communications management plan<br>• Stakeholder engagement plan<br>.3 Project documents updates<br>• Issue log<br>• Lessons learned register<br>• Project schedule<br>• Risk register<br>• Stakeholder register<br>.4 OPA updates |

## 10.2 Manage Communications

**Input:**

**01** Project management plan
- Resource management plan
- Communications management plan
- Stakeholder engagement plan

**02** Project documents
- Change log
- Issue log
- Lessons learned register
- Quality report
- Risk report
- Stakeholder register

**03** Work performance reports

**04** Enterprise environmental factors

**05** Organizational process assets

# 10.2 Manage Communications

## Tools & Techniques (1/2)

- **01** Communication technology
- **02** Communication methods
- **03** Communication skills
  - Communication competence
  - Feedback
  - Nonverbal
  - Presentations
- **04** Project management information system
- **05** Project reporting

## 10.2 Manage Communications

**Tools & Techniques (2/2)**

**06 Interpersonal and team skills**
- Active listening
- Conflict management
- Cultural awareness
- Meeting management
- Networking
- Political awareness

**07 Meetings**

## 10.2 Manage Communications

**Output:**

 **Project communications**

 **Project management plan updates**
- Communications management plan
- Stakeholder engagement plan

**Project documents updates**
- Issue log
- Lessons learned register
- Project schedule
- Risk register
- Stakeholder register

**Organizational process assets updates**

## 10.3 Monitor Communications

👍 **Is the process of** ensuring the information needs of the project and its stakeholders are met

🗝 **The key benefit** is the optimal information flow as defined in the communications management plan and the stakeholder engagement plan

## 10.3 Monitor Communications

### Inputs

.1 Project management plan
  - Resource management plan
  - Communications management plan
  - Stakeholder engagement plan

.2 Project documents
  - Issue log
  - Lessons learned register
  - Project communications

.3 Work performance data
.4 EEF
.5 OPA

### Tools & Techniques

.1 Expert judgment
.2 PMIS
.3 Data analysis
  - Stakeholder engagement assessment matrix
.4 Interpersonal and team skills
  - Observation/conversation
.5 Meetings

### Outputs

.1 Work performance information
.2 Change requests
.3 Project management plan updates
  - Communications management plan
  - Stakeholder engagement plan
.4 Project documents updates
  - Issue log
  - Lessons learned register
  - Stakeholder register

# 10.3 Monitor Communications

## Input:

**01** Project management plan
- Resource management plan
- Communications management plan
- Stakeholder engagement plan

**02** Project documents
- Issue log
- Lessons learned register
- Project communications

**03** Work performance data

**04** Enterprise environmental factors

**05** Organizational process assets

## 10.3 Monitor Communications

### Tools & Techniques (1/1)

01. Expert judgment
02. Project management information system
03. Data analysis
    - Stakeholder engagement assessment matrix
04. Interpersonal and team skills
    - Observation/conversation
05. Meetings

## 10.3 Monitor Communications

**Output:**

**01 Work performance information**

**02 Change requests**

**03 Project management plan updates**
- Communications management plan
- Stakeholder engagement plan

**04 Project documents updates**
- Issue log
- Lessons learned register
- Stakeholder register

**Have You Any question?**

| What | Why | When | Where | How | who |

# 11. PROJECT RISK MANAGEMENT

Presented by:
**Nasser Al Mohimeed**
PMO Director, ISO 21500 Lead Project Manager
Certified Project Managers Trainer

## Project Risk Management

**Project Risk Management** includes the processes of conducting risk management planning, identification, analysis, response planning, response implementation, and monitoring risk on a project.

**The objectives** of project risk management are to increase the probability and/or impact of positive risks and to decrease the probability and/or impact of negative risks, in order to optimize the chances of project success.

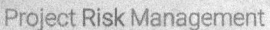

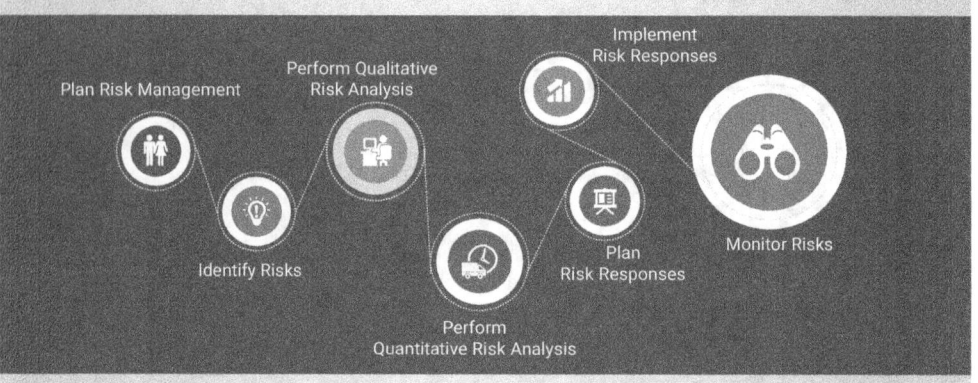

# Project Risk Management

| Initiating | Planning | Executing | Monitoring & Controlling | Closing |
|---|---|---|---|---|
| | 11.1 Plan Risk Management | 11.6 Implement Risk Responses | 11.7 Monitor Risks | |
| | 11.2 Identify Risks | | | |
| | 11.3 Perform Qualitative Risk Analysis | | | |
| | 11.4 Perform Quantitative Risk Analysis | | | |
| | 11.5 Plan Risk Responses | | | |

## Key concepts for Project Risk Management

- All projects are risky since they are unique undertakings with varying degrees of complexity that aim to deliver benefits.

- Project Risk Management aims to identify and manage risks that are not addressed by the other project management processes.

- The effectiveness of Project Risk Management is directly related to project success.

- Project Risk Management processes address two levels of risk in projects:
  - **Individual project risk** is an uncertain event or condition that, if it occurs, has a positive or negative effect on one or more project objectives.
  - **Overall project risk** is the effect of uncertainty on the project as a whole, arising from all sources of uncertainty including individual risks, representing the exposure of stakeholders to the implications of variations in project outcome, both positive and negative.

## Key concepts for Project Risk Management

- Project Risk Management aims to exploit or enhance positive risks (opportunities) while avoiding or mitigating negative risks (threats).

- Management of overall project risk aims to keep project risk exposure within an acceptable range by reducing drivers of negative variation, promoting drivers of positive variation, and maximizing the probability of achieving overall project objectives.

- Risks will continue to emerge during the lifetime of the project, so Project Risk Management processes should be conducted iteratively.

- Project team needs to know what level of risk exposure is acceptable in pursuit of the project objectives. ( Risk thresholds).

- Risk thresholds express the degree of acceptable variation around a project objective.

## TRENDS AND EMERGING PRACTICES IN PROJECT RISK MANAGEMENT

- **Non-event risks.**
  - **Variability risk.** Uncertainty exists about some key characteristics of a planned event or activity or decision.(productivity may be above or below target, the number of errors found during testing may be higher or lower than expected)
  - **Ambiguity risk.** Uncertainty exists about what might happen in the future. Areas of the project where imperfect knowledge might affect the project's ability to achieve its objectives include: (elements of the requirement or technical solution, future developments in regulatory frameworks, or inherent systemic complexity in the project).

  Can be addressed using Monte Carlo analysis.

- **Integrated risk management.** This builds risk efficiency into the structure of programs and portfolios, providing the greatest overall value for a given level of risk exposure.

## TRENDS AND EMERGING PRACTICES IN PROJECT RISK MANAGEMENT

- **Project resilience.** The existence of emergent risk is becoming clear, with a growing awareness of so-called unknowable-unknowns.
  These are risks that can only be recognized after they have occurred.
  Emergent risks can be tackled through developing project resilience. This requires :
  - Right level of budget and schedule contingency for emergent risks
  - Flexible project processes that can cope with emergent risk including strong change management;
  - Empowered project team that has clear objectives and that is trusted to get the job done Frequent review of early warning signs to identify emergent risks as early as possible;
  - Clear input from stakeholders to clarify areas where the project scope or strategy can be adjusted in response to emergent risks.

# TAILORING CONSIDERATIONS

- Because each project is unique, it is necessary to tailor the way Project Risk Management processes are applied.
- **Project size.** Does the project's size in terms of budget, duration, scope, or team size require a more detailed approach to risk management? Or is it small enough to justify a simplified risk process?
- **Project complexity.** Is a robust risk approach demanded by high levels of innovation, new technology, commercial arrangements, interfaces, or external dependencies that increase project complexity? Or is the project simple enough that a reduced risk process will suffice?
- **Project importance.** How strategically important is the project? Is the level of risk increased for this project because it aims to produce breakthrough opportunities, addresses significant blocks to organizational performance, or involves major product innovation?
- **Development approach.** Is this a waterfall project, where risk processes can be followed sequentially and iteratively, or does the project follow an agile approach where risk is addressed at the start of each iteration as well as during its execution?

## CONSIDERATIONS FOR AGILE/ADAPTIVE ENVIRONMENTS

- High-variability environments, by definition, incur more uncertainty and risk.
- PM using adaptive approaches to accelerate knowledge sharing and ensure that risk is understood and managed.
- Risk is considered when selecting the content of each iteration, and risks will also be identified, analyzed, and managed during each iteration.
- Work may be reprioritized as the project progresses, based on an improved understanding of current risk exposure.

## 11.1 Plan Risk Management

**PLAN RISK MANAGEMENT**
is the process of defining how to conduct risk management activities for a project

**THE KEY BENEFIT**
it ensures that the degree, type, and visibility of risk management are proportionate to both risks and the importance of the project to the organization and other stakeholders

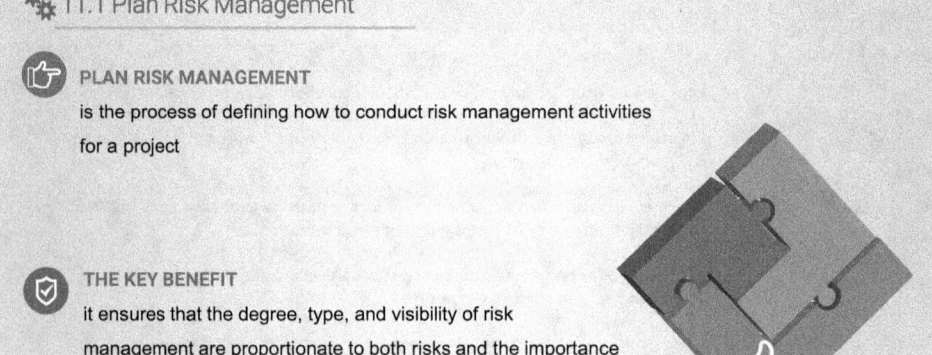

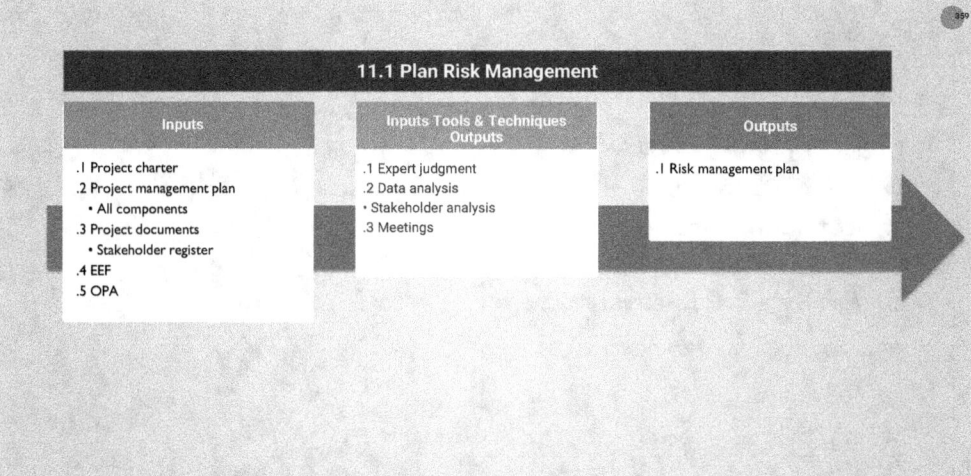

## 11.1 Plan Risk Management — Input

**01 PROJECT CHARTER**

**02 PROJECT MANAGEMENT PLAN**
- All components

**03 PROJECT DOCUMENTS**
- Stakeholder register

**04 ENTERPRISE ENVIRONMENTAL FACTORS**

**05 ORGANIZATIONAL PROCESS ASSETS**

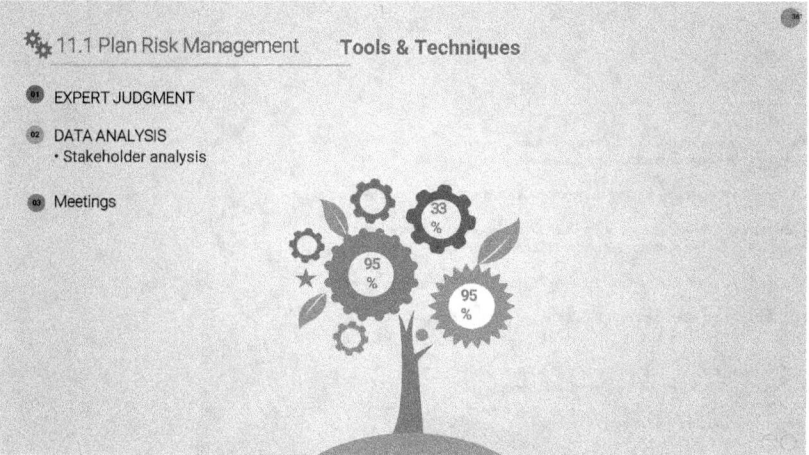

## 11.1 Plan Risk Management — Output 1

### RISK MANAGEMENT PLAN
Describes how risk management activities will be structured and performed.

- **Risk strategy.** Describes the general approach to managing risk on this project.

- **Methodology.** Defines the specific approaches, tools, and data sources that will be used.

- **Roles and responsibilities.** Defines the lead, support, and risk management team members for each type of activity described in the risk management plan, and clarifies their responsibilities.

- **Funding.** Identifies the funds needed to perform activities related to Project Risk Management.

- **Timing.** Defines when and how often the Project Risk Management processes will be performed

- **Risk categories.** Provide a means for grouping individual project risks.
Risk breakdown structure (RBS), which is a hierarchical representation of potential sources of risk
RBS helps the project team consider the full range of sources from which individual project risks may arise.

# 11.1 Plan Risk Management — Output 2

**Risk Breakdown Structure (RBS)**

| RBS LEVEL 0 | RBS LEVEL 1 | RBS LEVEL 2 |
|---|---|---|
| 0. ALL SOURCES OF PROJECT RISK | 1. TECHNICAL RISK | 1.1 Scope definition |
| | | 1.2 Requirements definition |
| | | 1.3 Estimates, assumptions, and constraints |
| | | 1.4 Technical processes |
| | | 1.5 Technology |
| | | 1.6 Technical interfaces |
| | | Etc. |
| | 2. MANAGEMENT RISK | 2.1 Project management |
| | | 2.2 Program/portfolio management |
| | | 2.3 Operations management |
| | | 2.4 Organization |
| | | 2.5 Resourcing |
| | | 2.6 Communication |
| | | Etc. |
| | 3. COMMERCIAL RISK | 3.1 Contractual terms and conditions |
| | | 3.2 Internal procurement |
| | | 3.3 Suppliers and vendors |
| | | 3.4 Subcontracts |
| | | 3.5 Client/customer stability |
| | | 3.6 Partnerships and joint ventures |
| | | Etc. |
| | 4. EXTERNAL RISK | 4.1 Legislation |
| | | 4.2 Exchange rates |
| | | 4.3 Site/facilities |
| | | 4.4 Environmental/weather |
| | | 4.5 Competition |
| | | 4.6 Regulatory |
| | | Etc. |

## 11.1 Plan Risk Management  Output 3

- **Stakeholder risk appetite.** The risk appetites of key stakeholders on the project are recorded in the risk management plan, as they inform the details of the Plan Risk Management process stakeholder risk appetite should be expressed as measurable risk thresholds

- **Definitions of risk probability and impacts.** used to evaluate both threats and opportunities by interpreting the impact definitions as negative for threats (delay, additional cost, and performance shortfall) and positive for opportunities (reduced time or cost, and performance enhancement).

| SCALE | PROBABILITY | +/- IMPACT ON PROJECT OBJECTIVES | | |
|---|---|---|---|---|
| | | TIME | COST | QUALITY |
| Very High | >70% | >6 months | >$5M | Very significant impact on overall functionality |
| High | 51-70% | 3-6 months | $1M-$5M | Significant impact on overall functionality |
| Medium | 31-50% | 1-3 months | $501K-$1M | Some impact in key functional areas |
| Low | 11-30% | 1-4 weeks | $100K-$500K | Minor impact on overall functionality |
| Very Low | 1-10% | 1 week | <$100K | Minor impact on secondary functions |
| Nil | <1% | No change | No change | No change in functionality |

## 11.1 Plan Risk Management  Output 4

- **Probability and impact matrix.** Opportunities and threats are represented in a common probability and impact matrix using positive definitions of impact for opportunities and negative impact definitions for threats. Descriptive terms
(such as very high, high, medium, low, and very low) or numeric values can be used for probability and impact. Where numeric values are used, these can be multiplied to give a probability-impact score for each risk.

| | | Threats | | | | | Opportunities | | | | |
|---|---|---|---|---|---|---|---|---|---|---|---|
| Very High 0.90 | 0.05 | 0.09 | 0.18 | 0.36 | 0.72 | 0.72 | 0.36 | 0.18 | 0.09 | 0.05 | Very High 0.90 |
| High 0.70 | 0.04 | 0.07 | 0.14 | 0.28 | 0.56 | 0.56 | 0.28 | 0.14 | 0.07 | 0.04 | High 0.70 |
| Medium 0.50 | 0.03 | 0.05 | 0.10 | 0.20 | 0.40 | 0.40 | 0.20 | 0.10 | 0.05 | 0.03 | Medium 0.50 |
| Low 0.30 | 0.02 | 0.03 | 0.06 | 0.12 | 0.24 | 0.24 | 0.12 | 0.06 | 0.03 | 0.02 | Low 0.30 |
| Very Low 0.10 | 0.01 | 0.01 | 0.02 | 0.04 | 0.08 | 0.08 | 0.04 | 0.02 | 0.01 | 0.01 | Very Low 0.10 |
| Probability | Very Low 0.05 | Low 0.10 | Moderate 0.20 | High 0.40 | Very High 0.80 | Very High 0.80 | High 0.40 | Moderate 0.20 | Low 0.10 | Very Low 0.05 | Probability |
| | | | Negative Impact | | | | | Positive Impact | | | |

## 11.1 Plan Risk Management — Output 5

- **Reporting formats.** Reporting formats define how the outcomes of the Project Risk Management process will be documented, analyzed, and communicated.

- **Tracking.** Tracking documents how risk activities will be recorded and how risk management processes will be audited.

## 11.2 Identify Risks

**IDENTIFY RISKS**
is the process of identifying individual project risks as well as sources of overall project risk, and documenting their characteristics.

**THE KEY BENEFIT**
the documentation of existing individual project risks and the sources of overall project risk. It also brings together information so the project team can respond appropriately to identified risks.

This process is performed throughout the project

## 11.2 Identify Risks

- Identify Risks considers both individual project risks and sources of overall project risk.

- identification activities may include the following: project manager, project team members, project risk specialist (if assigned), customers, subject matter experts from outside the project team, end users, other project managers, operations managers, stakeholders, and risk management experts within the organization.

- A consistent format should be used for risk statements to ensure that each risk is understood clearly

- Risk owners for individual project risks may be nominated as part of the Identify Risks process, and will be confirmed during the Perform Qualitative Risk Analysis process. Preliminary risk responses may also be identified and recorded and will be reviewed and confirmed as part of the Plan Risk Responses process.

- Identify Risks is an iterative process, The frequency of iteration and participation in each risk identification cycle will vary by situation, and this will be defined in the risk management plan.

## 11.2 Identify Risks

### Inputs

.1 Project management plan
  - Requirements management plan
  - Schedule management plan
  - Cost management plan
  - Quality management plan
  - Resource management plan
  - Risk management plan
  - Scope baseline
  - Schedule baseline
  - Cost baseline
.2 Project documents
  - Assumption log
  - Cost estimates
  - Duration estimates
  - Issue log
  - Lessons learned register
  - Requirements documentation
  - Resource requirements
  - Stakeholder register
.3 Agreements
.4 Procurement documentation
.5 EEF
.6 OPA

### Inputs Tools & Techniques Outputs

.1 Expert judgment
.2 Data gathering
  - Brainstorming
  - Checklists
  - Interviews
.3 Data analysis
  - Root cause analysis
  - Assumption and constraint analysis
  - SWOT analysis
  - Document analysis
.4 Interpersonal and team skills
  - Facilitation
.5 Prompt lists
.6 Meetings

### Outputs

.1 Risk register
.2 Risk report
.3 Project documents updates
  - Assumption log
  - Issue log
  - Lessons learned register

## 11.2 Identify Risks  Input

**01 PROJECT MANAGEMENT PLAN**
- Requirements management plan
- Schedule management plan
- Cost management plan
- Quality management plan
- Resource management plan
- Risk management plan
- Scope baseline
- Schedule baseline
- Cost baseline

**02 PROJECT DOCUMENTS**
- Assumption log
- Cost estimates
- Duration estimates
- Issue log
- Lessons learned register
- Requirements documentation
- Resource requirements
- Stakeholder register

**03 AGREEMENTS**

**04 PROCUREMENT DOCUMENTATION**

**05 ENTERPRISE ENVIRONMENTAL FACTORS**

**06 ORGANIZATIONAL PROCESS ASSETS**

## 11.2 Identify Risks — Tools & Techniques

**01 EXPERT JUDGMENT**

**02 DATA GATHERING**
- **Brainstorming.** to obtain a comprehensive list of individual project risks and sources of overall project risk.
- **Checklists.** developed based on historical information and knowledge that has been accumulated from similar projects and from other sources of information.
- **Interviews.** interviewing experienced participants, stakeholders, and subject matter experts to identify risks

**03 DATA ANALYSIS**
- **Root cause analysis.** to discover the underlying causes that lead to a problem, and develop preventive action.
- **Assumption and constraint analysis.** To explores the validity of assumptions and constraints to determine which pose a risk to the project
- **SWOT analysis.** examines the project from each of the strengths, weaknesses, opportunities, and threats (SWOT) perspectives.
- **Document analysis.**

## 11.2 Identify Risks    Tools & Techniques 2

**INTERPERSONAL AND TEAM SKILLS**

**PROMPT LISTS**
A prompt list is a predetermined list of risk categories that might give rise to individual project risks and that could also act as sources of overall project risk.
common strategic frameworks for identifying sources of overall project risk
- PESTLE (political, economic, social, technological, legal, environmental),
- TECOP (technical, environmental, commercial, operational, political), or
- VUCA (volatility, uncertainty, complexity, ambiguity).

**MEETINGS**
(often called a risk workshop).

## 11.2 Identify Risks

### ● RISK REGISTER
The risk register captures details of identified individual project risks. include :
- **List of identified risks.**
- **Potential risk owners.**
- **List of potential risk responses.**

### ● RISK REPORT
presents information on sources of overall project risk, together with summary information on identified individual project risks.
The risk report is developed progressively throughout the Project Risk Management process.
include but is not limited to:
- Sources of overall project risk,
- Summary information on identified individual project risks,

### ● PROJECT DOCUMENTS UPDATES
- Assumption log.
- Issue log..
- Lessons learned register

## 11.3 Perform Qualitative Risk Analysis

**Perform Quantitative Risk Analysis**
Is the process of prioritizing individual project risks for further analysis or action by assessing their probability of occurrence and impact as well as other characteristics

**THE KEY BENEFIT**
it focuses efforts on high-priority risks. This process is performed throughout the project

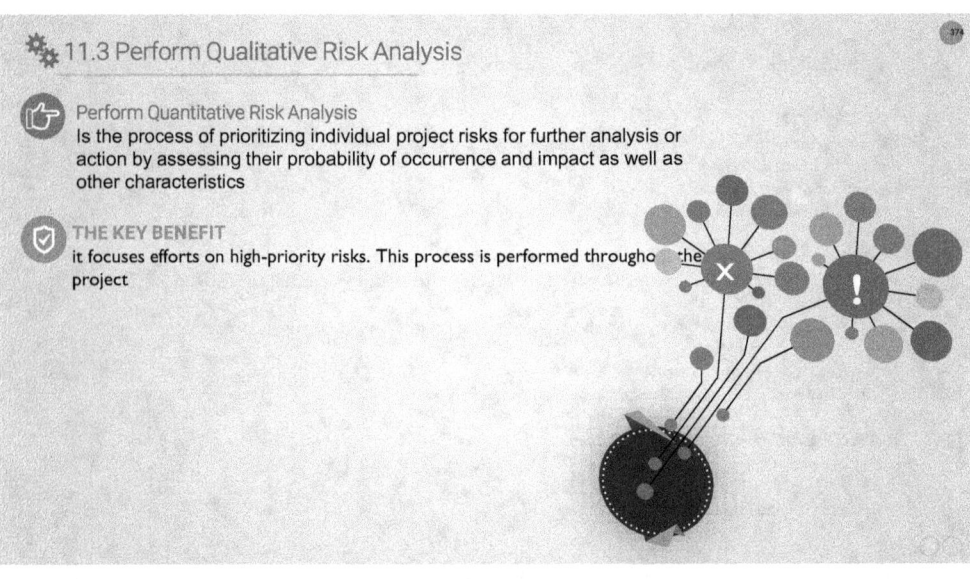

## 11.3 Perform Qualitative Risk Analysis

### Inputs
.1 Project management plan
  • Risk management plan
.2 Project documents
  • Assumption log
  • Risk register
  • Stakeholder register
.3 EEF
.4 OPA

### Inputs Tools & Techniques Outputs
.1 Expert judgment
.2 Data gathering
  • Interviews
.3 Data analysis
  • Risk data quality assessment
  • Risk probability and impact assessment
  • Assessment of other risk parameters
.4 Interpersonal and team skills
  • Facilitation
.5 Risk categorization
.6 Data representation
  • Probability and impact matrix
  • Hierarchical charts
.7 Meetings

### Outputs
.1 Project documents updates
  • Assumption log
  • Issue log
  • Risk register
  • Risk report

## 11.3 Perform Qualitative Risk Analysis — Input

**01 PROJECT MANAGEMENT PLAN**
- Risk management plan

**02 PROJECT DOCUMENTS**
- Assumption log
- Risk register
- Stakeholder register

**03 ENTERPRISE ENVIRONMENTAL FACTORS**

**04 ORGANIZATIONAL PROCESS ASSETS**

## 11.3 Perform Qualitative Risk Analysis     Tools & Techniques 1

**① EXPERT JUDGMENT**

**② DATA GATHERING**

**③ DATA ANALYSIS**
**Risk data quality assessment.** evaluates the degree to which the data about individual project risks is accurate and reliable as a basis for qualitative risk analysis.
- A weighted average of selected data quality characteristics can then be generated to
- give an overall quality score.

**Risk probability and impact assessment.** considers the likelihood that a specific risk will occur.
- Considers the potential effect project objectives ( schedule, cost, quality, or performance). Impacts will be negative for threats and positive for opportunities.

## 11.3 Perform Qualitative Risk Analysis    Tools & Techniques 2

**Assessment of other risk parameters.** consider other characteristics of risk, include:
- **Urgency**. The period of time within which a response to the risk is to be implemented in order to be effective.(A short period indicates high urgency.)
- **Proximity**. The period of time before the risk might have an impact on one or more project objectives. (A short period indicates high proximity.)
- **Dormancy**. The period of time that may elapse after a risk has occurred before its impact is discovered. (A short period indicates low dormancy.)
- **Manageability**. The ease with which the risk owner can manage the occurrence or impact of risk.
- **Controllability**. The degree to which the risk owner is able to control the risk's outcome.
- **Detectability**. The ease with which the results of the risk occurring, or being about to occur, can be detected and recognized.
- **Connectivity**. The extent to which the risk is related to other individual project risks.
- **Strategic impact**. The potential for the risk to have a positive or negative effect on the organization's strategic goals
- **Propinquity**. The degree to which a risk is perceived to matter by one or more stakeholders. Where a risk is perceived as very significant, propinquity is high.

04 **INTERPERSONAL AND TEAM SKILLS**

## 11.3 Perform Qualitative Risk Analysis — Tools & Techniques 3

**RISK CATEGORIZATION**
Risks to the project can be categorized by sources of risk (RBS); or other useful categories (e.g., project phase, project budget, and roles and responsibilities)

**DATA REPRESENTATION**
**Probability and impact matrix.**
**Hierarchical charts.** Where risks have been categorized using more than two parameters, Bubble chart displays three dimensions of data, where each risk is plotted (bubble), and the three parameters are represented by the x-axis value, the y-axis value, and the bubble size.

**MEETINGS**

## 11.3 Perform Qualitative Risk Analysis — Tools & Techniques 4

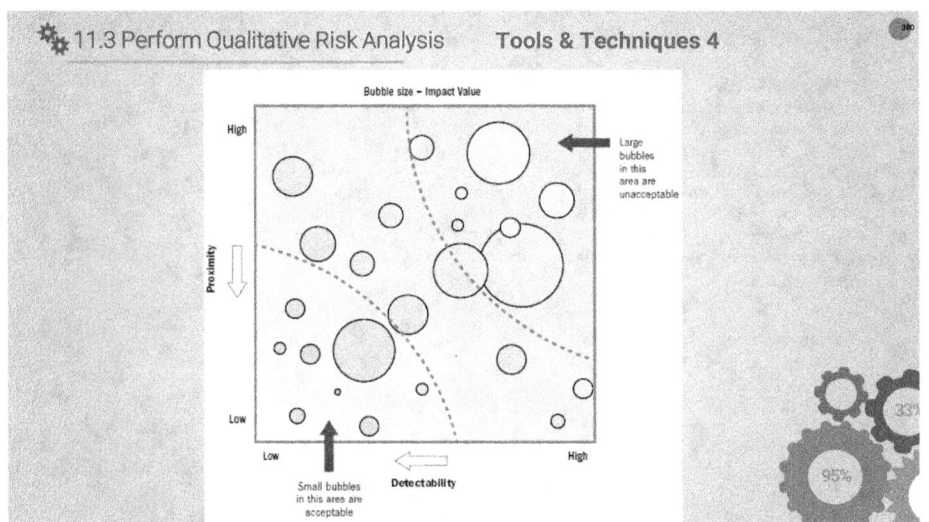

## 11.3 Perform Qualitative Risk Analysis — Output

**PROJECT DOCUMENTS UPDATES**
- Assumption log
- Issue log
- Risk register
- Risk report

## 11.4 Perform Quantitative Risk Analysis

**PERFORM QUANTITATIVE RISK ANALYSIS**
is the process of numerically analyzing the combined effect of identified individual project risks and other sources of uncertainty on overall project objectives

**THE KEY BENEFIT**
It quantifies overall project risk exposure, and it can also provide additional quantitative risk information to support risk response planning.

- This process is not required for every project, but where it is used, it is performed throughout the project

## 11.4 Perform Quantitative Risk Analysis

**Inputs**

.1 Project management plan
  • Risk management plan
  • Scope baseline
  • Schedule baseline
  • Cost baseline
.2 Project documents
  • Assumption log
  • Basis of estimates
  • Cost estimates
  • Cost forecasts
  • Duration estimates
  • Milestone list
  • Resource requirements
  • Risk register
  • Risk report
  • Schedule forecasts
.3 EEF
.4 OPA

**Inputs Tools & Techniques Outputs**

.1 Expert judgment
.2 Data gathering
  • Interviews
.3 Interpersonal and team skills
  • Facilitation
.4 Representations of uncertainty
.5 Data analysis
  • Simulations
  • Sensitivity analysis
  • Decision tree analysis
  • Influence diagrams

**Outputs**

.1 Project documents updates
  • Risk report

## 11.4 Perform Quantitative Risk Analysis — Input

**01 PROJECT MANAGEMENT PLAN**
- Risk management plan
- Scope baseline
- Schedule baseline
- Cost baseline

**02 PROJECT DOCUMENTS**
- Assumption log
- Basis of estimates
- Cost estimates
- Cost forecasts
- Duration estimates
- Milestone list
- Resource requirements
- Risk register
- Risk report
- Schedule forecasts

**03 ENTERPRISE ENVIRONMENTAL FACTORS**

**04 ORGANIZATIONAL PROCESS ASSETS**

## 11.4 Perform Quantitative Risk Analysis — Tools & Techniques 1

**01 EXPERT JUDGMENT**

**02 DATA GATHERING**

**03 INTERPERSONAL AND TEAM SKILLS**

**04 REPRESENTATIONS OF UNCERTAINTY**
This may take several forms. triangular, normal, lognormal, beta, uniform, or discrete distributions.
Care should be taken when selecting an appropriate probability distribution to reflect the range of possible values for the planned activity.

**05 DATA ANALYSIS**
**Influence diagrams.** represents a project or situation within the project as a set of entities, outcomes, and influences, together with the relationships and effects between them
Outputs from an influence diagram are similar to other quantitative risk analysis methods, including S-curves and tornado diagrams.

**Simulation.** Simulates the combined effects of individual project risks and other sources of uncert... evaluate their potential impact on achieving project objectives
performed using a Monte Carlo analysis.

## 11.4 Perform Quantitative Risk Analysis — Tools & Techniques 2

**Monte Carlo analysis**
When running a Monte Carlo analysis for cost risk, the simulation uses the project cost estimates.
When running a Monte Carlo analysis for schedule risk, the schedule network diagram and duration estimates are used.

Computer software is used to iterate the quantitative risk analysis model. The input values (e.g., cost estimates, duration estimates, or occurrence of probabilistic branches)

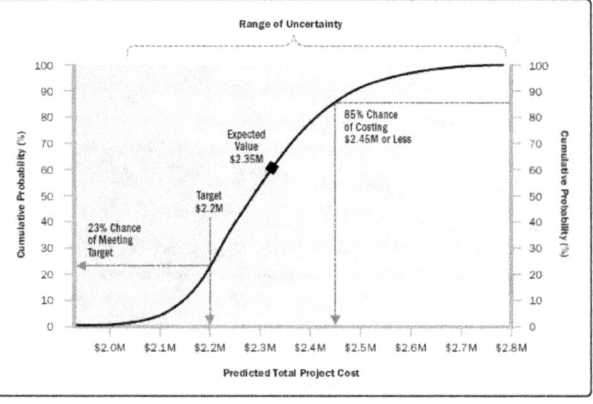

Figure 11-13. Example S-Curve from Quantitative Cost Risk Analysis

## 11.4 Perform Quantitative Risk Analysis — Tools & Techniques 3

**Sensitivity analysis.** helps to determine which individual project risks or other sources of uncertainty have the most potential impact on project outcomes.

Tornado diagram, which presents the calculated correlation coefficient for each element of the quantitative risk analysis model that can influence the project outcome.

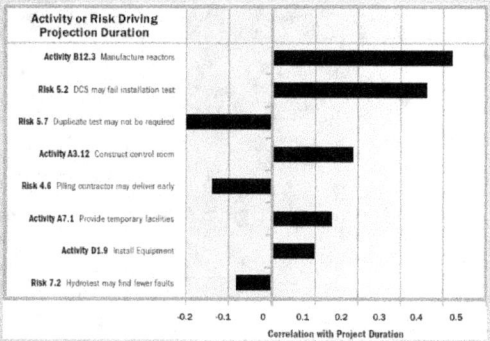

## 11.4 Perform Quantitative Risk Analysis — Tools & Techniques 4

**Decision tree analysis.** used to support selection of the best of several alternative courses of action.

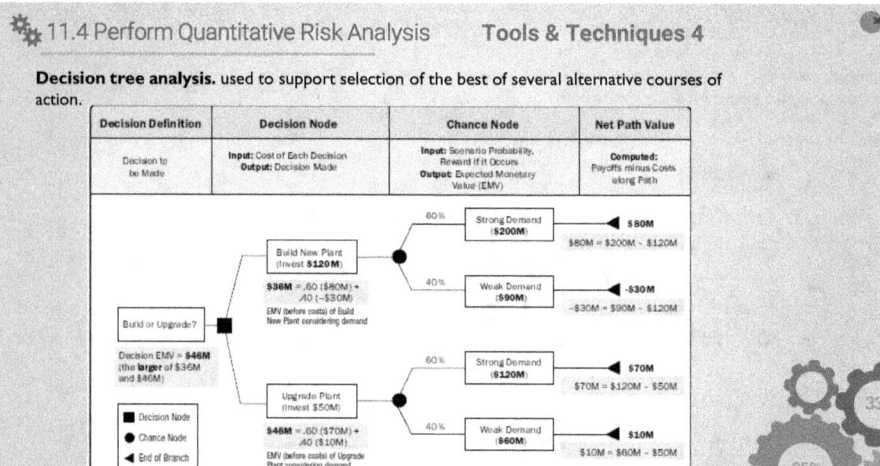

## 11.4 Perform Quantitative Risk Analysis — Output

- **PROJECT DOCUMENTS UPDATES**
  - **Assessment of overall project risk exposure**
    - Chances of project success
    - Degree of inherent variability remaining within the project at the time the analysis was conducted.
  - **Detailed probabilistic analysis of the project.**
    - Amount of contingency reserve needed to provide a specified level of confidence;
    - Identification of individual project risks or other sources of uncertainty that have the greatest effect on the project critical path;
    - Major drivers of overall project risk, with the greatest influence on uncertainty in project outcomes.

  - **Prioritized list of individual project risks.**

  - **Trends in quantitative risk analysis results.**

  - **Recommended risk responses.**

## 11.5 Plan Risk Responses

**PLAN RISK RESPONSES**
is the process of developing options, selecting strategies, and agreeing on actions to address overall project risk exposure, as well as to treat individual project risks

**THE KEY BENEFIT**
It identifies appropriate ways to address overall project risk and individual project risks.
- This process also allocates resources and inserts activities into project documents and the project management plan as needed.
- This process is performed throughout the project.

## 11.5 Plan Risk Responses

**Inputs**
- .1 Project management plan
  - Resource management plan
  - Risk management plan
  - Cost baseline
- .2 Project documents
  - Lessons learned register
  - Project schedule
  - Project team assignments
  - Resource calendars
  - Risk register
  - Risk report
  - Stakeholder register
- .3 EEF
- .4 OPA

**Tools & Techniques**
- .1 Expert judgment
- .2 Data gathering
  - Interviews
- .3 Interpersonal and team skills
  - Facilitation
- .4 Strategies for threats
- .5 Strategies for opportunities
- .6 Contingent response strategies
- .7 Strategies for overall project risk
- .8 Data analysis
  - Alternatives analysis
  - Cost-benefit analysis
- .9 Decision making
  - Multicriteria decision analysis

**Outputs**
- .1 Change requests
- .2 Project management plan updates
  - Schedule management plan
  - Cost management plan
  - Quality management plan
  - Resource management plan
  - Procurement management plan
  - Scope baseline
  - Schedule baseline
  - Cost baseline
- .3 Project documents updates
  - Assumption log
  - Cost forecasts
  - Lessons learned register
  - Project schedule
  - Project team assignments
  - Risk register
  - Risk report

## 11.5 Plan Risk Responses — Input

**01 PROJECT MANAGEMENT PLAN**
- Resource management plan
- Risk management plan
- Cost baseline

**02 PROJECT DOCUMENTS**
- Lessons learned register
- Project schedule
- Project team assignments
- Resource calendars
- Risk register
- Risk report
- Stakeholder register

**03 ENTERPRISE ENVIRONMENTAL FACTORS**

**04 ORGANIZATIONAL PROCESS ASSETS**

## 11.5 Plan Risk Responses — Tools & Techniques 1

- **EXPERT JUDGMENT**
- **DATA GATHERING**
- **INTERPERSONAL AND TEAM SKILLS**
- **STRATEGIES FOR THREATS**
    1. **Escalate.** appropriate when a threat is outside the scope of the project or that the proposed response would exceed the project manager's authority.
    PM determines who should be notified about the threat and communicates the details to it

    2. **Avoid.** when the project team acts to eliminate the threat or protect the project from its impact. appropriate for high-priority threats with a high probability of occurrence and a large negative impact.

    3. **Transfer.** shifting ownership of a threat to a third party to manage the risk

    4. **Mitigate.** action is taken to reduce the probability of occurrence and/or impact of a threat.

    5. **Accept.** acknowledges the existence of a threat, but no proactive action is taken, appropriate for low-priority threats.

## 11.5 Plan Risk Responses — Tools & Techniques 2

### STRATEGIES FOR OPPORTUNITIES
1. **Escalate.**

2. **Exploit.** The exploit strategy may be selected for high-priority opportunities where the organization wants to ensure that the opportunity is realized.
   by ensuring that it definitely happens, increasing the probability of occurrence to 100%

3. **Share.** Sharing involves transferring ownership of an opportunity to a third party so that it shares some of the benefit if the opportunity occurs.

4. **Enhance.** used to increase the probability and/or impact of an opportunity.

5. **Accept.** Accepting an opportunity acknowledges its existence but no proactive action is taken. This strategy may be appropriate for low-priority opportunities.

### CONTINGENT RESPONSE STRATEGIES
It is appropriate for the project team to make a response plan that will only be executed under certain predefined conditions

## 11.5 Plan Risk Responses — Tools & Techniques 3

**STRATEGIES FOR OVERALL PROJECT RISK**
1. Avoid
2. Exploit.
3. Transfer/share.
4. Mitigate/enhance.
5. Accept

**DATA ANALYSIS**
- Alternatives analysis
- Cost-benefit analysis.

**DECISION MAKING**

## 11.5 Plan Risk Responses — Output

**01 CHANGE REQUESTS**

**02 PROJECT MANAGEMENT PLAN UPDATES**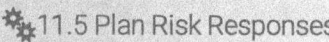
- Schedule management plan
- Cost management plan
- Quality management plan
- Resource management plan
- Procurement management plan
- Scope baseline
- Schedule baseline
- Cost baseline

**03 PROJECT DOCUMENTS UPDATES**
- Assumption log
- Cost forecasts
- Lessons learned register
- Project schedule
- Project team assignments
- Risk register
- Risk report

## 11.6 Implement Risk Responses

**IMPLEMENT RISK RESPONSES**
is the process of implementing agreed-upon risk response plans.

**THE KEY BENEFIT**
it ensures that agreed-upon risk responses are executed as planned in order to address overall project risk exposure, minimize individual project threats, and maximize individual project opportunities.

- This process is performed throughout the project.

## 11.6 Implement Risk Responses

**Inputs**
- .1 Project management plan
  - Risk management plan
- .2 Project documents
  - Lessons learned register
  - Risk register
  - Risk report
- .3 OPA

**Tools & Techniques**
- .1 Expert judgment
- .2 Interpersonal and team skills
  - Influencing
- .3 PMIS

**Outputs**
- .1 Change requests
- .2 Project documents updates
  - Issue log
  - Lessons learned register
  - Project team assignments
  - Risk register
  - Risk report

## 11.6 Implement Risk Responses — Input

**01 PROJECT MANAGEMENT PLAN**
- Risk management plan

**02 PROJECT DOCUMENTS**
- Lessons learned register
- Risk register
- Risk report

**03 ORGANIZATIONAL PROCESS ASSETS**

## 11.6 Implement Risk Responses — Tools & Techniques 1

**01** **EXPERT JUDGMENT**

**02** **INTERPERSONAL AND TEAM SKILLS**
 • Influencing

**03** **PROJECT MANAGEMENT INFORMATION SYSTEM**

## 11.6 Implement Risk Responses — Output

**① CHANGE REQUESTS**

**② PROJECT DOCUMENTS UPDATES**
- Issue log
- Lessons learned register
- Project team assignments
- Risk register
- Risk report

## 11.7 Monitor Risks

**MONITOR RISKS**
is the process of monitoring the implementation of agreed-upon risk response plans, tracking identified risks, identifying and analyzing new risks, and evaluating risk process effectiveness throughout the project.

**THE KEY BENEFIT**
it enables project decisions to be based on current information about overall project risk exposure and individual project risks.

- This process is performed throughout the project.

## 11.7 Monitor Risks

**The Monitor Risks process uses performance information generated during project execution to determine if:**
- Implemented risk responses are effective,
- Level of overall project risk has changed,
- Status of identified individual project risks has changed,
- New individual project risks have arisen,
- Risk management approach is still appropriate,
- Project assumptions are still valid,
- Risk management policies and procedures are being followed,
- Contingency reserves for cost or schedule require modification, and
- Project strategy is still valid.

## 11.7 Monitor Risks

**Inputs**
- .1 Project management plan
  - Risk management plan
- .2 Project documents
  - Issue log
  - Lessons learned register
  - Risk register
  - Risk report
- .3 Work performance data
- .4 Work performance reports

**Inputs Tools & Techniques Outputs**
- .1 Data analysis
  - Technical performance analysis
  - Reserve analysis
- .2 Audits
- .3 Meetings

**Outputs**
- .1 Work performance information
- .2 Change requests
- .3 Project management plan updates
  - Any component
- .4 Project documents updates
  - Assumption log
  - Issue log
  - Lessons learned register
  - Risk register
  - Risk report
- .5 OPA updates

## 11.7 Monitor Risks — Input

**01 PROJECT MANAGEMENT PLAN**
- Risk management plan

**02 PROJECT DOCUMENTS**
- Issue log
- Lessons learned register
- Risk register
- Risk report

**03 WORK PERFORMANCE DATA**

**04 WORK PERFORMANCE REPORTS**

## 11.7 Monitor Risks — Tools & Techniques 1

**01 DATA ANALYSIS**
- **Technical performance analysis.** compares technical accomplishments during project execution to the schedule of technical achievement
- **Reserve analysis.** compares the amount of the contingency reserves remaining to the amount of risk remaining to determine if the remaining reserve is adequate

**02 AUDITS**

**03 MEETINGS**

## 11.7 Monitor Risks  Output

**01  WORK PERFORMANCE INFORMATION**
Change requests

**02  PROJECT MANAGEMENT PLAN UPDATES**
- Any component

**03  PROJECT DOCUMENTS UPDATES**
- Assumption log
- Issue log
- Lessons learned register
- Risk register
- Risk report

**04  ORGANIZATIONAL PROCESS ASSETS UPDATES**

# 12. PROJECT
## PROCUREMENT MANAGEMENT

Presented by :
**Nasser Al Mohimeed**
PMO Director, ISO 21500 Lead Project Manager
Certified Project Managers Trainer

# PROJECT PROCUREMENT MANAGEMENT

**Project Procurement Management** includes the processes necessary to purchase or acquire products, services, or results needed from outside the project team

**Project Procurement.** Includes the management and control processes required to develop and administer agreements such as contracts, purchase orders, memoranda of agreements (MOAs), or internal service level agreements (SLAs).

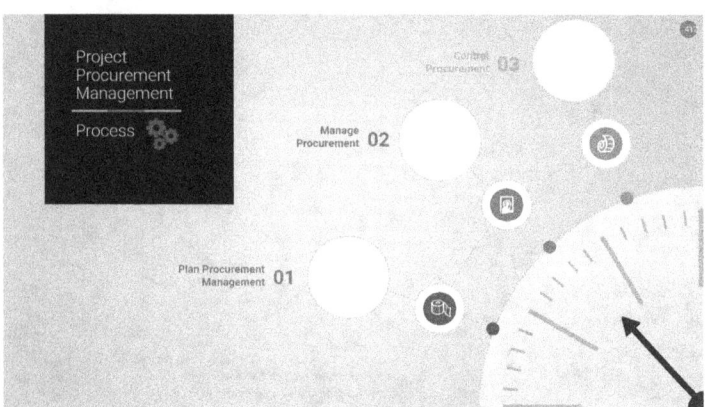

## Project Procurement Management

| Initiating | Planning | Executing | Monitoring & Controlling | Closing |
|---|---|---|---|---|
| | 12.1 Plan Procurement Management | 12.2 Control Procurement | 12.3 Control Procurement | |

## Key concepts for Project Procurement Management

- The Project Procurement Management processes involve agreements that describe the relationship between two parties—a buyer and a seller. Agreements can be as simple or complex as multiyear international construction contracts.

- A contract should clearly state the deliverables and results expected, including any knowledge transfer from the seller to the buyer.

- A contract means it will be subjected to a more extensive approval process, often involving the legal department

- A complex project may involve managing multiple contracts simultaneously or in sequence

- Seller may be identified as a contractor, vendor, service provider, or supplier.

- The buyer may be the owner of the final product, a subcontractor, the acquiring organization, a service requestor, or the purchaser.

## Key concepts for Project Procurement Management

- The winning bidder may manage the work as a project. In such cases:
  - The buyer becomes the customer to subcontractors, suppliers, and service providers
  - The seller's project management team may be concerned with all the processes involved in performing the work or providing the services.
  - Terms and conditions of the contract and the procurement statement of work (SOW) become key inputs to many of the seller's management processes.
  - The seller itself may become a buyer of lower-tiered products, services, and materials from subcontractors and suppliers.

## Trends and emerging practices in Procurement Management

Advances in tools
- **Online tools** for procurement give the buyers a single point where procurements can be advertised and provide sellers with a single source to find procurement documents and complete them directly online.
- Using **building information model** (BIM) in software tools to save significant amounts of time and money on projects using it.
- This approach can substantially reduce construction claims, thereby reducing both costs and schedule.

More advanced risk management
- Write contracts that accurately allocate **specific risks** to those entities most capable of managing them.
- No contractor is capable of managing all the possible **major risks** on a project.
- The buyer will be required to **accept the risks** that the contractors do not have control over.
- Contracts may specify that risk management be performed as **part of the contract**.

Changing contracting processes.
- There has been a significant growth in **mega-projects**
- The contractor works closely with the client in the procurement process to take **advantage of discounts** through quantity purchases or other special considerations.
- For these projects, the use of internationally recognized standard contract forms is increasing in order to **reduce problems and claims** during execution

## Trends and emerging practices in Procurement Management

Logistics and supply chain management.
- **Long-lead items** may be procured in advance of other procurement contracts to meet the planned project completion date.
- It is possible to begin contracting for these **long-lead materials, supplies, or equipment** before the final design of the end product itself is completed based on **the known requirements** identified in the top-level design.

Technology and stakeholder relations.
- Use of technology including webcams to improve **stakeholder communications** and relations.
- The **progress** on the project can be viewed on the Internet by all stakeholders.
- Video data can also be stored, allowing analysis if a claim arises.

Trial engagements.
- some projects will engage several **candidate sellers** for initial deliverables and work products on a paid basis before making the full commitment to a larger portion of the project scope.
- This **accelerates momentum** by allowing the buyer to **evaluate** potential partners, while simultaneously making progress on project work.

## Key concepts for Project Procurement Management

**TAILORING CONSIDERATIONS**

- Complexity of procurement. Is there one main procurement or are there multiple procurements at different times with different sellers that add to the complexity of the procurements?
- Physical location Are the buyers and sellers in the same location, or reasonably close, or in different time zones, countries, or continents?
- Governance and regulatory environment. Are local laws and regulations regarding procurement activities integrated with the organization's procurement policies? How does this affect contract auditing requirements?
- Availability of contractors. Are there available contractors who are capable of performing the work?

## Key concepts for Project Procurement Management

### CONSIDERATIONS FOR AGILE/ADAPTIVE ENVIRONMENTS

- Specific sellers may be used to extend the team.
- This collaborative working relationship can lead to a shared risk procurement model where both the buyer and the seller share in the risk and rewards associated with a project.

In large projects,
- May use an adaptive approach for some deliverables and a more stable approach for other parts.
- In these cases, a governing agreement such as a master services agreement (MSA) may be used for the overall engagement, with the adaptive work being placed in an appendix or supplement.
- This allows changes to occur on the adaptive scope without impacting the overall contract.

## 12.1 Plan Procurement Management

**PLAN PROCUREMENT MANAGEMENT**
the process of documenting project procurement decisions, specifying the approach and identifying potential sellers.

**THE KEY BENEFIT**
is that it determines whether to acquire goods and services from outside the project and, if so, what to acquire as well as how and when to acquire it. Goods and services may be procured from other parts of the performing organization or from external sources.

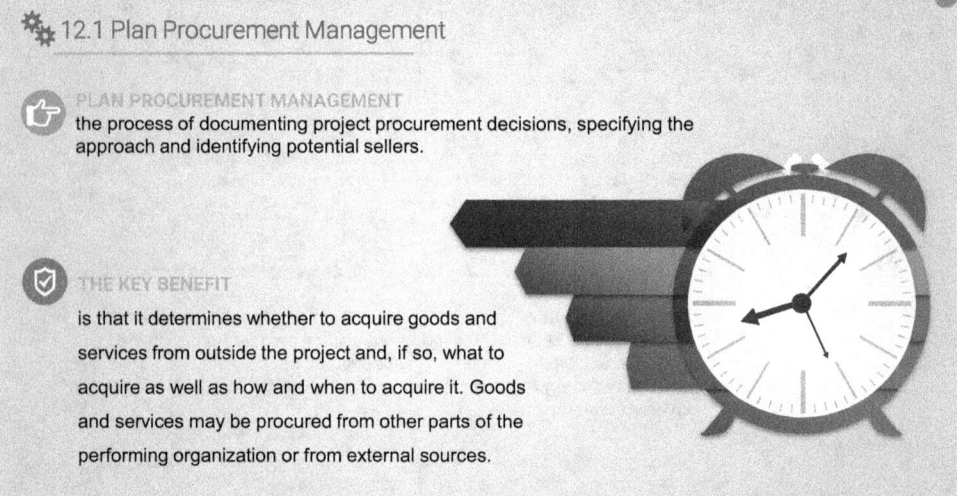

## 12.1 Plan Procurement Management

### Input
.1 Project charter
.2 Business documents
  • Business case
  • Benefits management plan
.3 Project management plan
  • Scope management plan
  • Quality management plan
  • Resource management plan
  • Scope baseline
.4 Project documents
  • Milestone list
  • Project team assignments
  • Requirements documentation
  • Requirements traceability matrix
  • Resource requirements
  • Risk register
  • Stakeholder register
.5 EEF
.6 OPA

### Inputs Tools & Techniques Outputs
.1 Expert judgment
.2 Data gathering
  • Market research
.3 Data analysis
  • Make-or-buy analysis
.4 Source selection analysis
.5 Meetings

### Outputs
.1 Procurement management plan
.2 Procurement strategy
.3 Bid documents
.4 Procurement statement of work
.5 Source selection criteria
.6 Make-or-buy decisions
.7 Independent cost estimates
.8 Change requests
.9 Project documents updates
  • Lessons learned register
  • Milestone list
  • Requirement documentation
  • Requirements traceability matrix
  • Risk register
  • Stakeholder register
.10 OPA updates

## 12.1 Plan Procurement Management

**Typical steps might be:**
1. Prepare the procurement statement of work (SOW) or terms of reference (TOR).
2. Prepare a high-level cost estimate to determine the budget.
3. Advertise the opportunity.
4. Identify a short list of qualified sellers.
5. Prepare and issue bid documents.
6. Prepare and submit proposals by the seller.
7. Conduct a technical evaluation of the proposals including quality.
8. Perform a cost evaluation of the proposals.
9. Prepare the final combined quality and cost evaluation to select the winning proposal.
10. Finalize negotiations and sign contract between the buyer and the seller.

## 12.1 Plan Procurement Management — Input

**01 PROJECT CHARTER**

**02 BUSINESS DOCUMENTS**
- Business case
- Benefits management plan

**03 PROJECT MANAGEMENT PLAN**
- Scope management plan
- Quality management plan
- Resource management plan
- Scope baseline.

**04 PROJECT MANAGEMENT PLAN.**
- Milestone list.
- Project team assignments.
- Requirements documentation.
- Requirements traceability matrix.
- Resource requirements.
- Risk register.
- Stakeholder register.

**ENTERPRISE ENVIRONMENTAL FACTORS**

## 12.1 Plan Procurement Management — Input 2

 **ORGANIZATIONAL PROCESS ASSETS**
Can influence the Plan Procurement Management process include but are not limited to:
- **Preapproved seller lists**
- **Formal procurement policies, procedures, and guidelines.**
- **Contract types:** fixed-price, cost-reimbursable, or time and materials contract.

Fixed-price contracts.
- Setting a fixed total price for a defined product, service, or result to be provided.
- Should be used when the requirements are well defined and no significant changes to the scope are expected.

➢ **Firm fixed price (FFP).** Commonly used and favored by most buyer because the price for goods is set at the outset and not subject to change unless the scope of work changes.

➢ **Fixed price incentive fee (FPIF).** Gives the buyer and seller some flexibility in that it allows for deviation from performance, with financial incentives tied to achieving agreed-upon metrics. A price ceiling is set, and all costs above the price ceiling are the responsibility of the seller.

➢ **Fixed price with economic price adjustments (FPEPA).** It is a fixed-price contract, with a special provision allowing adjustments to the contract price due to changed conditions, such as inflation changes

**12.1 Plan Procurement Management    Input 2**

Cost-reimbursable contracts.
- Involves payments (cost reimbursements) to the seller for all actual costs incurred for completed work, plus a fee representing seller profit.
- Used if the scope of work is expected to change significantly during the execution of the contract.

➢ **Cost plus fixed fee (CPFF).** The seller is reimbursed for all allowable costs for performing the contract work and receives a fixed-fee payment calculated as a percentage of the initial estimated project costs.

➢ **Cost plus incentive fee (CPIF).** The seller is reimbursed for all allowable costs for performing the contract work and receives a predetermined incentive fee based on achieving certain performance objectives
In CPIF contracts, if the final costs are less or greater than the original estimated costs, then both the buyer and seller share costs based on pre negotiated cost-sharing formula,

➢ **Cost plus award fee (CPAF).** Seller is reimbursed for all legitimate costs, but the majority of the fee is earned based on the satisfaction of certain performance criteria defined into the contract.

Time and material contracts (T&M). (also called time and means) are a hybrid type of contractual arrangement with aspects of both cost-reimbursable and fixed-price contracts.

## 1٢.1 Plan Procurement Management    Tools & Techniques

- **EXPERT JUDGMENT**

- **DATA GATHERING**
  Market research includes examination of industry and specific seller capabilities.

- **DATA ANALYSIS**
  **Make-or-buy analysis**
  - is used to determine whether work or deliverables can best be accomplished by the project team or should be purchased from outside sources.
  - Factors to consider include the organization's current resource allocation and their skills and abilities, the need for specialized expertise, the desire to not expand permanent employment obligations, and the need for independent expertise.
  - It also includes evaluating the risks involved with each make-or-buy decision.
  - Make-or-buy analysis may use payback period, return on investment (ROI), internal rate of return (IRR), discounted cash flow, net present value (NPV), benefit/cost analysis (BCA), or other techniques in order to decide whether to include something as part of the project or purchase it externally.

## ١٢.1 Plan Procurement Management    Tools & Techniques

**04 SOURCE SELECTION ANALYSIS**

**Least cost.**
- Appropriate for procurements of a standard or routine nature where well-established practices and standards exist and from which a specific and well-defined outcome is expected, which can be executed at different costs.

**Qualifications only.**
- Applies when the time and cost of a full selection process would not make sense because the value of the procurement is relatively small.
- The buyer establishes a short list and selects the bidder with the best credibility, qualifications, experience, expertise, areas of specialization, and references.

**Quality-based/highest technical proposal score.**
The selected firm is asked to submit a proposal with both technical and cost details technical proposals are evaluated based on the quality of the technical solution offered. The seller who submitted the highest-ranked technical proposal is selected if their financial proposal can be negotiated and accepted.

## ۱۲.1 Plan Procurement Management — Tools & Techniques

**Quality and cost-based.**
- Allows cost to be included as a factor in the seller selection process.
- When risk are greater for project, quality should be a key element when compared to cost.

**Sole source.**
- The buyer asks a specific seller to prepare technical and financial proposals,
- There is no competition,

**Fixed budget.**
- In the RFP and selecting the highest-ranking technical proposal within the budget.
- Because the cost constraint, seller will adapt the scope and quality of their offer to that budget.
- The buyer should ensure that the budget is compatible with the SOW.
- This method is appropriate only when the SOW is precisely defined, no changes are anticipated.

 **MEETINGS**

## 12.1 Plan Procurement Management  Output 1

**PROCUREMENT MANAGEMENT PLAN**
- It contains the activities to be undertaken during the procurement process.
- IT should document whether international competitive bidding, national competitive bidding, local bidding, etc., should be done.
- If the project is financed externally, the sources and availability of funding should be aligned with the procurement management plan and the project schedule.
- It may be formal or informal, highly detailed, or broadly framed based on the needs of the project, and includes appropriate control thresholds.
- The Procurement management plan can establish the following:
    - How procurement will be coordinated with other project aspects.
    - Timetable of key procurement activities.
    - Procurement metrics to be used to manage contracts.
    - Stakeholder roles and responsibilities related to procurement.
    - Constraints and assumptions that could affect planned procurements.
    - The legal jurisdiction and the currency in which payments will be made.
    - Risk management issues.
    - Prequalified sellers, if any, to be used.

## 12.1 Plan Procurement Management — Output 2

**PROCUREMENT STRATEGY**

The objective of the procurement strategy is to determine:

1. **Delivery methods.**
   - For professional services, delivery methods include:
     buyer/services provider with no subcontracting, buyer/services provider with subcontracting allowed, joint venture between buyer and services provider, or buyer/ services provider acts as the representative.
   - For industrial or commercial construction, project delivery methods include but are not limited to: turnkey, design build (DB), design bid build (DBB), design build operate (DBO), build own operate transfer (BOOT), and others.

2. **Contract payment types.** lump sum, firm fixed price, cost plus award fees, cost plus incentive fees, time and materials, target cost, and others.

3. **Procurement phases.**
   - Sequencing or phasing of the procurement, a description of each phase
   - Procurement performance indicators and milestones to be used in monitoring;
   - Criteria for moving from phase to phase;
   - Monitoring and evaluation plan for tracking progress; and
   - Process for knowledge transfer for use in subsequent phases.

## 12.1 Plan Procurement Management — Output 3

- **BID DOCUMENTS**
  - Used to solicit proposals from prospective sellers.
  - Terms such as bid, tender, or quotation are used when the seller selection decision is based on price
  - Term such as proposal is used when other considerations such as technical capability or technical approach are the most important.
  - Biding document can include:
    - **Request for information (RFI):** is used when more information on the goods and services to be acquired, needed from the sellers.
    - **Request for quotation (RFQ)** : used when more information is needed on how vendors would satisfy the requirements and/or how much it will cost.
    - **Request for proposal (RFP).** is used when there is a problem in the project and the solution is not easy to determine.

## 12.1 Plan Procurement Management — Output 4

### PROCUREMENT STATEMENT OF WORK (SOW)
- Developed from the project scope baseline and defines only that portion of the to be included inn contract.
- The SOW describes the procurement item in sufficient detail (specifications, quantity desired, quality levels, performance data, period of performance, work location, and other requirements).
- The procurement SOW should be clear, complete, and concise.
- terms of reference (TOR) is sometimes used when contracting for services.
- SOW, a TOR typically includes these elements:
  - Tasks the contractor is required to perform as well as specified coordination requirements;
  - Standards the contractor will fulfill that are applicable to the project;
  - Data that needs to be submitted for approval;
  - Detailed list of all data and services that will be provided to the contractor by the buyer
  - Definition of the schedule for initial submission and the review/approval time required.

### SOURCE SELECTION CRITERIA

## 12.1 Plan Procurement Management — Output 5

- **MAKE-OR-BUY DECISIONS**

- **INDEPENDENT COST ESTIMATES**
  - For large procurements, the procuring organization may elect to either prepare its own independent estimate to serve as a benchmark on proposed responses.
  - The specific criteria may be a numerical score, color-code, or a written description of how well the seller satisfies the buying organization's needs.
  - The criteria will be part of a weighting system that can be used to select a single seller that will be asked to sign a contract and establish a negotiating sequence by ranking all the proposals by the weighted evaluation scores assigned to each proposal.

- **CHANGE REQUESTS**

- **PROJECT DOCUMENTS UPDATES.**
  - Lessons learned register.
  - Milestone list.
  - Requirements documentation.
  - Requirements traceability matrix.
  - Risk register.
  - Stakeholder register.

- **ORGANIZATIONAL PROCESS ASSETS UPDATES**

# 12.1 Plan Procurement Management

Table 12-1. Comparison of Procurement Documentation

| Procurement Management Plan | Procurement Strategy | Statement of Work | Bid Documents |
|---|---|---|---|
| How procurement work will be coordinated and integrated with other project work, particularly with resources, schedule, and budget | Procurement delivery methods | Description of the procurement item | Request for information (RFI), Request for quote (RFQ), Request for proposal (RFP) |
| Timetable for key procurement activities | Type of agreements | Specifications, quality requirements and performance metrics | |
| Procurement metrics to manage the contract | Procurement phases | Description of collateral services required | |
| Responsibilities of all stakeholders | | Acceptance methods and criteria | |
| Procurement assumptions and constraints | | Performance data and other reports required | |
| Legal jurisdiction and currency used for payment | | Quality | |
| Information on independent estimates | | Period and place of performance | |
| Risk management issues | | Currency; payment schedule | |
| Prequalified sellers, if applicable | | Warranty | |

## 12.2 Conduct Procurement Management

 **CONDUCT PROCUREMENT MANAGEMENT**
the process of obtaining seller responses, selecting a seller, and awarding a contract.

 **THE KEY BENEFIT**
is that it selects a qualified seller and implements the legal agreement for delivery. The end results of the process are the established agreements including formal contracts.

## 12.2 Conduct Procurement Management

### Input

.1 Project management plan
- Scope management plan
- Requirements management plan
- Communications management plan
- Risk management plan
- Procurement management plan
- Configuration management plan
- Cost baseline

.2 Project documents
- Lessons learned register
- Project schedule
- Requirements documentation
- Risk register
- Stakeholder register

.3 Procurement documentation
.4 Seller proposals
.5 EEF
.6 OPA

### Inputs Tools & Techniques Outputs

.1 Expert judgment
.2 Advertising
.3 Bidder conferences
.4 Data analysis
- Proposal evaluation

.5 Interpersonal and team skills
- Negotiation

### Outputs

.1 Selected sellers
.2 Agreements
.3 Change requests
.4 Project management plan updates
- Requirements management plan
- Quality management plan
- Communications management plan
- Risk management plan
- Procurement management plan
- Scope baseline
- Schedule baseline
- Cost baseline

.5 Project documents updates
- Lessons learned register
- Requirements documentation
- Requirements traceability matrix
- Resource calendars
- Risk register
- Stakeholder register

.6 OPA updates

## 12.2 Conduct Procurement Management — Input - 1

 **PROJECT MANAGEMENT PLAN**
- Scope management plan
- Requirements management plan
- Communications management plan
- Risk management plan
- Procurement management plan
- Configuration management plan
- Cost baseline

**PROJECT DOCUMENTS**
- Lessons learned register.
- Project schedule.
- Requirements documentation.
- Risk register.
- Stakeholder register.

## 12.2 Conduct Procurement Management — Input - 2

**PROCUREMENT DOCUMENTATION**
- Bid documents.
- Procurement statement of work.
- Independent cost estimates.
- Source selection criteria.

**SELLER PROPOSALS**
- If the seller is going to submit a price proposal, good practice is to require that it be separate from the technical proposal.
- The evaluation body reviews each submitted proposal according to the source selection criteria and selects the seller that can best satisfy the buying organization's requirements.

**ENTERPRISE ENVIRONMENTAL FACTORS(EEF).**

**ORGANIZATIONAL PROCESS ASSETS (OPA).**

## 12.2 Conduct Procurement Management — Tools & Techniques - 1

- **EXPERT JUDGMENT**

- **ADVERTISING**
  - Advertising is communicating with potential users of a product, service, or result.
  - Existing lists of potential sellers often can be expanded by placing advertisements in general circulation publications such as selected newspapers or in specialty trade publications.
  - Most government jurisdictions require public advertising or online posting of pending government contracts.

- **BIDDER CONFERENCES**
  - (also called contractor conferences, vendor conferences, and pre-bid conferences) are meetings between the buyer and prospective sellers prior to proposal submittal.
  - They are used to ensure that all prospective bidders have a clear and common understanding of the procurement and no bidders receive preferential treatment.

## 12.2 Conduct Procurement Management — Tools & Techniques - 2

- **DATA ANALYSIS**
  - **Proposal evaluation.** to ensure they are complete and respond in full to the bid documents, procurement statement of work, source selection criteria, and any other documents that went out in the bid package.

-  **INTERPERSONAL AND TEAM SKILLS**
  **Negotiation**
  - Procurement negotiation clarifies the structure, rights, and obligations of the parties and other terms of the purchases so that mutual agreement can be reached prior to signing the contract.
  - Final document language reflects all agreements reached.
  - Negotiation concludes with a signed contract document or other formal agreement that can be executed by both buyer and seller.
  - The negotiation should be led by a member of the procurement team that has the authority to sign contracts.
  - The project manager and other members of the project management team may be present during negotiation to provide assistance as needed.

**12.2 Conduct Procurement Management** — Outputs - 1

**SELECTED SELLERS**
- The selected sellers are those who have been judged to be in a competitive range based on the outcome of the proposal or bid evaluation.
- Final approval of complex, high-value, high-risk
- procurements will generally require organizational
- senior management approval prior to award.

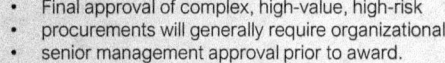

## 12.2 Conduct Procurement Management — Outputs - 2

### AGREEMENTS

A contract is a mutually binding agreement that obligates the seller to provide the specified products, services, or results; obligates the buyer to compensate the seller; and represents a legal relationship that is subject to remedy in the courts.

The major components in an agreement document :
- Procurement statement of work or major deliverables;
- Schedule, milestones, or date by which a schedule is required;
- Performance reporting;
- Pricing and payment terms;
- Inspection, quality, and acceptance criteria;
- Warranty and future product support;
- Incentives and penalties;
- Insurance and performance bonds;
- Subordinate subcontractor approvals;
- General terms and conditions;
- Change request handling; and
- Termination clause and alternative dispute resolution mechanisms.

### CHANGE REQUESTS.

## 12.2 Conduct Procurement Management — Outputs - 3

**PROJECT MANAGEMENT PLAN UPDATES**
- Requirements management plan.
- Quality management plan.
- Communications management plan.
- Risk management plan.
- Procurement management plan.
- Scope baseline.
- Schedule baseline.
- Cost baseline.

**PROJECT DOCUMENTS UPDATES**
- Lessons learned register.
- Requirements documentation.
- Requirements traceability matrix.
- Resource calendars.
- Risk register.
- Stakeholder register.

**ORGANIZATIONAL PROCESS ASSETS UPDATES**

## 12.3 CONTROL PROCUREMENTS

**Control Procurements**
Is the process of managing procurement relationships; monitoring contract performance, and making changes and corrections as appropriate; and closing out contracts.

**THE KEY BENEFIT**
that it ensures that both the seller's and buyer's performance meet the project's requirements according to the
terms of the legal agreement.
This process is performed throughout the project as needed.

##  12.3 CONTROL PROCUREMENTS

 Control Procurements includes application of the appropriate project management processes to the contractual relationship(s) and integration of the outputs from these processes into the overall management of the project.
This integration often occurs at multiple levels when there are multiple sellers and multiple products, services, or results involved.
- Administrative activities may include:
- Collection of data and managing project records,
- Refinement of procurement plans and schedules.
- Set up for gathering, analyzing, and reporting procurement-related project data and preparation of periodic reports to the organization.
- Monitoring the procurement environment so that implementation can be facilitated or adjustments made.
- Payment of invoices.

## 12.3 Control Procurement Management

**Input**

Project management plan
- Requirements management plan
- Risk management plan
- Procurement management plan
- Change management plan
- Schedule baseline

.2 Project documents
- Assumption log
- Lessons learned register
- Milestone list
- Quality reports
- Requirements documentation
- Requirements traceability matrix
- Risk register
- Stakeholder register

.3 Agreements
.4 Procurement documentation
.5 Approved change requests
.6 Work performance data
.7 EEF
.8 OPA

**Inputs Tools & Techniques Outputs**

.1 Expert judgment
.2 Claims administration
.3 Data analysis
- Performance reviews
- Earned value analysis
- Trend analysis

.4 Inspection
.5 Audits

**Outputs**

.1 Closed procurements
.2 Work performance information
.3 Procurement documentation updates
.4 Change requests
.5 Project management plan updates
- Risk management plan
- Procurement management plan
- Schedule baseline
- Cost baseline

.6 Project documents updates
- Lessons learned register
- Resource requirements
- Requirements traceability matrix
- Risk register
- Stakeholder register

.7 OPA updates

## 12.3 Control Procurement Management — Input - 1

**01 PROJECT MANAGEMENT PLAN**
- Requirements management plan.
- Risk management plan.
- Procurement management plan.
- Change management plan.
- Schedule baseline.

**02 PROJECT DOCUMENTS**
- Assumption log.
- Lessons learned register.
- Milestone list.
- Quality reports.
- Requirements documentation.
- Requirements traceability matrix.
- Risk register.
- Stakeholder register.

**03 AGREEMENTS**

## 12.3 Control Procurement Management — Input - 2

- **04** PROCUREMENT DOCUMENTATION
- **05** APPROVED CHANGE REQUESTS
- **06** WORK PERFORMANCE DATA
- **07** ENTERPRISE ENVIRONMENTAL FACTORS
- **08** ORGANIZATIONAL PROCESS ASSETS

## 12.2 Control Procurement Management — Tools & Techniques - 1

**EXPERT JUDGMENT**

**CLAIMS ADMINISTRATION**
- The contested changes are called claims. When they cannot be resolved, they become disputes and finally appeals.
- Claims are documented, processed, monitored, and managed throughout the contract life cycle, usually in accordance with the terms of the contract.
- If the parties themselves do not resolve a claim, it may have to be handled in accordance with alternative dispute resolution (ADR) typically following procedures established in the contract.
- Settlement of all claims and disputes through negotiation is the preferred method.

**DATA ANALYSIS**
- **Performance Reviews.** Measure, compare, and analyze quality, resource, schedule, and cost performance against the agreement.
- **Earned Value Analysis (EVA).**
- **Trend Analysis.** Trend analysis can develop a forecast estimate at completion (EAC)

## 12.3 Control Procurement Management — Tools & Techniques - 2

**INSPECTION**
a structured review of the work being performed by the contractor.

**AUDITS**
- A structured review of the procurement process.
- Rights and obligations related to audits should be described in the procurement contract.
- Resulting audit observations should be brought to the attention of the buyer's project manager and the seller's project manager for adjustments to the project, when necessary.

## 12.3 Control Procurement Management — Outputs - 1

**CLOSED PROCUREMENTS**
- The buyer, usually through its authorized procurement administrator, provides the seller with formal written notice that the contract has been completed.
- Requirements for formal procurement closure are usually defined in the terms and conditions of the contract and are included in the procurement management plan.
- Typically, all deliverables should have been provided on time and meet technical and quality requirements, there should be no outstanding claims or invoices, and all final payments should have been made.
- The project management team should have approved all deliverables prior to closure.

**WORK PERFORMANCE INFORMATION**
how a seller is performing by comparing the deliverables received, the technical performance achieved, and the costs incurred and accepted against the SOW budget for the work performed.

## 12.3 Control Procurement Management — Outputs - 2

**PROCUREMENT DOCUMENTATION UPDATES**
- includes the contract with all supporting schedules, requested unapproved contract changes, and approved change requests.
- Procurement documentation also includes any seller-developed technical documentation and other work performance information such as deliverables, seller performance reports and warranties, financial documents including invoices and payment records, and the results of contract-related inspections.

**PROJECT MANAGEMENT PLAN UPDATES**
- Risk management plan.
- Procurement management plan.
- Cost baseline.
- Schedule baseline.

## 12.3 Control Procurement Management — Outputs - 3

**PROJECT DOCUMENTS UPDATES**
- Lessons learned register.
- Resource requirements.
- Requirements traceability matrix.
- Risk register.
- Stakeholder register.

**ORGANIZATIONAL PROCESS ASSETS UPDATES**
- Payment schedules and requests.
- Seller performance evaluation documentation
- Prequalified seller lists updates.
- Lessons learned repository.
- Procurement file.

# 13. PROJECT
## STAKEHOLDER MANAGEMENT

Presented by :
**Nasser Al Mohimeed**
PMO Director, ISO 21500 Lead Project Manager
Certified Project Managers Trainer

# PROJECT STAKEHOLDER MANAGEMENT

**Project Stakeholder Management** includes the processes required to identify the people, groups, or organizations that could impact or be impacted by the project, to analyze stakeholder expectations and their impact on the project, and to develop appropriate management strategies for effectively engaging stakeholders in project decisions and execution.

# Project Stakeholder Management Process

**Identify Stakeholders** — 1

**Plan Stakeholder Engagement** — 2

**Manage Stakeholder Engagement** — 3

**Monitor Stakeholder Engagement** — 4

# Project Stakeholder Management

| Initiating | Planning | Executing | Monitoring & Controlling | Closing |
|---|---|---|---|---|
| 13.1 Identify Stakeholders | 13.2 Plan Stakeholder Engagement | 13.3 Manage Stakeholder Engagement | 13.4 Monitor Stakeholder Engagement | |

## Key concepts for Project stakeholder Management

- Every project has stakeholders who are impacted by or can impact the project in a positive or negative way.
- Some stakeholders may have a limited ability to influence the project's work or outcomes; others may have significant influence on the project and its expected outcomes.
- The ability of the project manager and team to correctly identify and engage all stakeholders in an appropriate way can mean the difference between project success and failure.
- Stakeholder satisfaction should be identified and managed as a project objective.
- The process of identifying and engaging stakeholders for the benefit of the project is iterative, should be reviewed and updated routinely
  - The project moves through different phases in its life cycle,
  - Current stakeholders are no longer involved in the work or new stakeholders become members of the project
  - There are significant changes in the organization or the wider stakeholder community.

## Trends and emerging practices in Stakeholder Management

- Identifying all stakeholders, not just a limited set.
- Ensuring that all team members are involved in stakeholder engagement activities.
- Reviewing the stakeholder community regularly, often in parallel with reviews of individual project risks.
- Consulting with stakeholders who are most affected by the work or outcomes of the project through the concept of co-creation.
- Co-creation places greater emphasis on including affected stakeholders in the team as partners.
- Capturing the value of effective stakeholder engagement, both positive and negative.

  > Positive value can be based on the consideration of benefits derived from higher levels of active support from stakeholders, particularly powerful stakeholders.

  > Negative value can be derived by measuring the true costs of not engaging stakeholders effectively, leading to product recalls or loss of organizational or project reputation.

## Key concepts for Project Stakeholder Management

### TAILORING CONSIDERATIONS

Because Each project is unique; therefore, the project manager will need to tailor the way Project stakeholder Management processes are applied

- ✓ **Stakeholder diversity.** How many stakeholders are there? How diverse is the culture within the stakeholder community?

- ✓ **Complexity of stakeholder relationships.** How complex are the relationships within the stakeholder community? The more networks a stakeholder or stakeholder group participates in, the more complex the networks of information and misinformation the stakeholder may receive.

- ✓ **Communication technology.** What communication technology is available? What support mechanisms are in place to ensure that best value is achieved from the technology?

## CONSIDERATIONS FOR AGILE/ADAPTIVE ENVIRONMENTS

- The high degree of change require active engagement and participation with project stakeholders.

- To facilitate timely, productive discussion and decision making, adaptive teams engage with stakeholders directly rather than going through layers of management.

- Often the client, user, and developer exchange information in a dynamic Co-creative process that leads to more stakeholder involvement and higher satisfaction.

- Regular interactions with the stakeholder community throughout the project mitigate risk, build trust, and support adjustments earlier in the project cycle, thus reducing costs and increasing the likelihood of success for the project.

- In order to accelerate the sharing of information within and across the organization, agile methods promote aggressive transparency.

- The intent of inviting any stakeholders to project meetings and reviews or posting project artifacts in public spaces is to surface as quickly as possible any misalignment, dependency, or other issue related to the changing project.\

## 13.1 Identify Stakeholders

**IDENTIFY STAKEHOLDERS**
the process of identifying project stakeholders regularly and analyzing and documenting relevant information regarding their interests, involvement, interdependencies, influence, and potential impact on project success.

**THE KEY BENEFIT**
is that it enables the project team to identify the appropriate focus for engagement of each stakeholder or group of stakeholders.

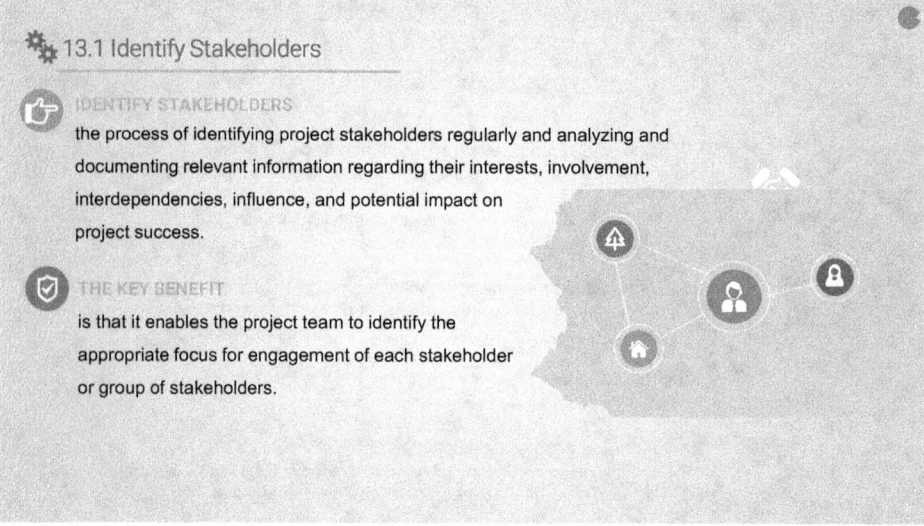

## 13.1 Identify Stakeholders

### Input

1. Project charter
.2 Business documents
  - Business case
  - Benefits management plan
.3 Project management plan
  - Communications management plan
  - Stakeholder engagement plan
.4 Project documents
  - Change log
  - Issue log
  - Requirements documentation
.5 Agreements
.6 EEF
.7 OPA

### Inputs Tools & Techniques Outputs

.1 Expert judgment
.2 Data gathering
  - Questionnaires and surveys
  - Brainstorming
.3 Data analysis
  - Stakeholder analysis
  - Document analysis
.4 Data representation
  - Stakeholder mapping/ representation
.5 Meetings

### Outputs

.1 Stakeholder register
.2 Change requests
.3 Project management plan updates
  - Requirements management plan
  - Communications management plan
  - Risk management plan
  - Stakeholder engagement plan
.4 Project documents updates
  - Assumption log
  - Issue log
  - Risk register

## 13.1 Identify Stakeholders — Input

**01** PROJECT CHARTER

**02** BUSINESS DOCUMENTS
- Business case.
- Benefits management plan.

**03** PROJECT MANAGEMENT PLAN
- Communications management plan.
- Stakeholder engagement plan.

**04** PROJECT DOCUMENTS
- Change log
- Issue log
- Requirements documentation

**05** AGREEMENTS
**06** ENTERPRISE ENVIRONMENTAL FACTORS
**07** ORGANIZATIONAL PROCESS ASSETS

## 13.1 Identify Stakeholders — Tools & Techniques

**01** Expert judgment.

**02** DATA GATHERING.

- ❖ **Questionnaires and surveys**. Can include one-on-one, reviews, focus group sessions, or other mass information collection techniques.

- ❖ **Brainstorming**. used to identify stakeholders can include both brainstorming and brain writing.
  - Brainstorming. A general data-gathering and creativity technique that elicits input from groups such as team members or subject matter experts.
  - Brain writing. A refinement of brainstorming that allows individual participants time to consider the question(s) individually before the group creativity session is held. The information can be gathered in face-to-face groups or using virtual environments supported by technology.

## 13.1 Identify Stakeholders — Tools & Techniques

**DATA ANALYSIS**
- Stakeholder analysis.

It results in a list of stakeholders and relevant information such as their positions in the organization, roles on the project, "stakes," expectations, attitudes (their levels of support for the project), and their interest in information about the project.

- Interest. A person or group can be affected by a decision related to the project or its outcomes.
- Rights (legal or moral rights). Legal rights, such as occupational health and safety, may be defined in the legislation framework of a country. Moral rights may involve concepts of protection of historical sites or environmental sustainability.
- Ownership. A person or group has a legal title to an asset or a property.
- Knowledge. Specialist knowledge, which can benefit the project through more effective delivery of project objectives, organizational outcomes, or knowledge of the power structures of the organization.
- Contribution. Provision of funds or other resources, including human resources, or providing support for the project in more intangible ways, such as advocacy in the form of promoting the objectives of the project or acting as a buffer between the project and the power structures of the organization and its politics.

- Document analysis. Assessing the available project documentation and lessons learned from previous projects to identify stakeholders and other supporting information.

## 13.1 Identify Stakeholders — Tools & Techniques

 **DATA REPRESENTATION**
❖ Stakeholder mapping/ representation.
A method of categorizing stakeholders using various methods.to assists the team in building relationships with the identified project stakeholders.
Include:

1. Power/interest grid, power/influence grid, or impact/influence grid.
   - Group stakeholders according to their level of authority (power),
     level of concern about the project's outcomes (interest),
     ability to influence the outcomes of the project (influence),
     or ability to cause changes to the project's planning or execution.
   - These classification models are useful for small projects or for projects with simple relationships between stakeholders and the project,
     or within the stakeholder community itself.

## 13.1 Identify Stakeholders     Tools & Techniques

2. Stakeholder cube.
- This is a refinement of the grid models previously mentioned.
- Combines (power, interest, and influence) into a three-dimensional model that can be useful in identifying and engaging their stakeholder community.
- It provides a model with multiple dimensions that improves the depiction of the stakeholder

3. Salience model.
- Describes classes of stakeholders based on assessments of their power, influence, <u>urgency</u> (need for immediate attention, either time-constrained or relating to the stakeholders' high stake in the outcome), and <u>legitimacy</u> (their involvement is appropriate).

- There is an adaptation of the salience model that substitutes proximity for legitimacy (applying to the team and measuring their level of involvement with the work of the project).

- The salience model is useful for large complex communities of stakeholders or where there are complex networks of relationships within the community.

- It is also useful in determining the relative importance of the identified stakeholders.

## 13.1 Identify Stakeholders — Tools & Techniques

4. **Directions of influence.**

Classifies stakeholders according to their influence on the work of the project or the project team itself. As the following :

- Upward: senior management of the performing organization or customer organization, sponsor, and steering committee.
- Downward: the team or specialists contributing knowledge or skills in a temporary capacity.
- Outward: stakeholder groups and their representatives outside the project team, such as suppliers, government departments, the public, end-users, and regulators.
- Sideward: the peers of the project manager, such as other project managers or middle managers who are in competition for scarce project resources or who collaborate with the project manager in sharing resources or information.

5. **Prioritization.**

necessary for projects with a large number of stakeholders, where the membership of the stakeholder community is changing frequently, or when the relationships between stakeholders and the project team or within the stakeholder community are complex.

**MEETINGS**

**13.1 Identify Stakeholders** **Output**

**① STAKEHOLDER REGISTER**
This document contains information about identified stakeholders
- Identification information. Name, organizational position, location and contact details, and role on the project.
- Assessment information. Major requirements, expectations, potential for influencing project outcomes, and the phase of the project life cycle where the stakeholder has the most influence or impact.
- Stakeholder classification. Internal/external, impact/influence/power/interest, upward/downward/outward/ sideward, or any other classification model chosen by the project manager.

**② CHANGE REQUESTS**

## 13.1 Identify Stakeholders — Output

### PROJECT MANAGEMENT PLAN UPDATES

- ✓ Requirements management plan.
- ✓ Communications management plan.
- ✓ Risk management plan.
- ✓ Stakeholder engagement plan.

### PROJECT DOCUMENTS UPDATES

- ✓ Assumption log.
- ✓ Issue log.
- ✓ Risk register.

##  13.2 Plan Stakeholder Engagement

 **PLAN STAKEHOLDER ENGAGEMENT**
the process of developing approaches to involve project stakeholders based on their needs, expectations, interests, and potential impact on the project.

 **THE KEY BENEFIT**
The key benefit is that it provides an actionable plan to interact effectively with stakeholders.

## 13.2 Plan Stakeholder Engagement

An effective plan that recognizes the diverse information needs of the project's stakeholders, developed early in the project life cycle, is reviewed and updated regularly as the stakeholder community changes.
- The stakeholder engagement plan is updated regularly to reflect changes to the stakeholder community.
- Typical trigger situations requiring updates to the plan include :
  - When it is the start of a new phase of the project;
  - When there are changes to the organization structure or within the industry;
  - When new individuals or groups become stakeholders, current stakeholders are no longer part of the stakeholder community, or the importance of particular stakeholders to the project's success changes
  - When outputs of other project process areas, such as change management, risk management, or issue management, require a review of stakeholder engagement strategies.
- The results of these adjustments may be changes to the relative importance of the stakeholders who have been identified.

## 13.2 Plan Stakeholder Engagement

### Inputs
.1 Project charter
.2 Project management plan
  • Resource management plan
  • Communications management plan
  • Risk management plan
.3 Project documents
  • Assumption log
  • Change log
  • Issue log
  • Project schedule
  • Risk register
  • Stakeholder register
.4 Agreements
.5 EEFs
.6 OPA

### Inputs Tools & Techniques Outputs
.1 Expert judgment
.2 Data gathering
  • Benchmarking
.3 Data analysis
  • Assumption and constraint analysis
  • Root cause analysis
.4 Decision making
  • Prioritization/ranking
.5 Data representation
  • Mind mapping
  • Stakeholder engagement assessment matrix
.6 Meetings

### Outputs
.1 Stakeholder engagement plan

## 13.2 Plan Stakeholder Engagement — Input

**01** **PROJECT CHARTER**

**02** **PROJECT MANAGEMENT PLAN**
- Resource management plan
- Communications management plan
- Risk management plan

**03** **Project documents**
- Assumption log
- Change log
- Issue log
- Project schedule
- Risk register
- Stakeholder register

# 13.2 Plan Stakeholder Engagement    Input

**AGREEMENTS**
When planning for the engagement of contractors and suppliers, coordination usually involves working with the procurement/contracting group in the organization to ensure contractors and suppliers are effectively managed.

**ENTERPRISE ENVIRONMENTAL FACTORS**

**ORGANIZATIONAL PROCESS ASSETS**

## 13.2 Plan Stakeholder Engagement — Tools & Techniques

- **Expert judgment**

- **DATA GATHERING**
  Benchmarking. The results of stakeholder analysis are compared with information from other organizations or other projects that are considered to be world class.

- **DATA ANALYSIS**
  - Assumption and constraint analysis : Analysis of current assumptions and constraints may be conducted in order to tailor appropriate engagement strategies.
  - Root cause analysis : identifies underlying reasons for the level of support of project stakeholders in order to select the appropriate strategy to improve their level of engagement.

- **DECISION MAKING**
  - Prioritization/Ranking techniques : Stakeholder requirements need to be prioritized and ranked, as do the stakeholders themselves. Stakeholders with the most interest and the highest influence are often prioritized at the top of the list.

## 13.2 Plan Stakeholder Engagement — Tools & Techniques

**DATA REPRESENTATION**
- **Mind mapping**
- **Stakeholder engagement assessment matrix.** supports comparison between the current engagement levels of stakeholders and the desired engagement levels required for successful project delivery.

The engagement level of stakeholders can be classified as follows:
- Unaware. Unaware of the project and potential impacts.
- Resistant. Aware of the project and potential impacts but resistant to any change that may occur as a result of the work or outcomes of the project. These stakeholders will be unsupportive of the work or outcomes of the project.
- Neutral. Aware of the project, but neither supportive nor unsupportive.
- Supportive. Aware of the project and potential impacts and supportive of the work and its outcomes.
- Leading. Aware of the project and potential impacts and actively engaged in ensuring that the project is a success.

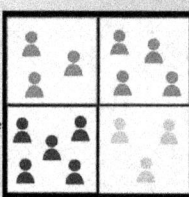

## 13.2 Plan Stakeholder Engagement — Tools & Techniques

| Stakeholder | Unaware | Resistant | Neutral | Supportive | Leading |
|---|---|---|---|---|---|
| Stakeholder 1 | C | | | D | |
| Stakeholder 2 | | | C | D | |
| Stakeholder 3 | | | | D C | |

Figure 13-6. Stakeholder Engagement Assessment Matrix

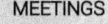 MEETINGS

**13.2 PLAN STAKEHOLDER ENGAGEMENT**  Output

**STAKEHOLDER ENGAGEMENT PLAN**
- ✓ A component of the project management plan that identifies the strategies and actions required to promote productive involvement of stakeholders in decision making and execution.

- ✓ It can be formal or informal and highly detailed or broadly framed, based on the needs of the project and the expectations of stakeholders.

- ✓ The stakeholder engagement plan may include but is not limited to specific strategies or approaches for engaging with individuals or groups of stakeholders.

# 13.3 MANAGE STAKEHOLDER ENGAGEMENT

 **MANAGE STAKEHOLDER ENGAGEMENT**
is the process of communicating and working with stakeholders to meet their needs and expectations, address issues, and foster appropriate stakeholder involvement.

**THE KEY BENEFIT**
it allows the project manager to increase support and minimize resistance from stakeholders.

## 13.3 MANAGE STAKEHOLDER ENGAGEMENT

Manage Stakeholder Engagement involves activities such as:
- Engaging stakeholders at appropriate project stages to obtain, confirm, or maintain their continued commitment to the success of the project.
- Managing stakeholder expectations through negotiation and communication;
- Addressing any risks or potential concerns related to stakeholder management and anticipating future issues that may be raised by stakeholders.
- Clarifying and resolving issues that have been identified.

✓ Managing stakeholder engagement helps to ensure that stakeholders clearly understand the project goals, objectives benefits, and risks for the project, as well as how their contribution will enhance project success.

## 13.3 MANAGE STAKEHOLDER ENGAGEMENT

### Inputs

.1 Project management plan
  - Communications management plan
  - Risk management plan
  - Stakeholder engagement plan
  - Change management plan
.2 Project documents
  - Change log
  - Issue log
  - Lessons learned register
  - Stakeholder register
.3 EEFs
.4 OPA

### Tools & Techniques

.1 Expert judgment
.2 Communication skills
  - Feedback
.3 Interpersonal and team skills
  - Conflict management
  - Cultural awareness
  - Negotiation
  - Observation/conversation
  - Political awareness
.4 Ground rules
.5 Meetings

### Outputs

.1 Change requests
.2 Project management plan updates
  - Communications management plan
  - Stakeholder engagement plan
.3 Project documents updates
  - Change log
  - Issue log
  - Lessons learned register
  - Stakeholder register

## 13.3 MANAGE STAKEHOLDER ENGAGEMENT — Input

**01 PROJECT MANAGEMENT PLAN**
- ✓ Communications management plan.
- ✓ Risk management plan.
- ✓ Stakeholder engagement plan.
- ✓ Change management plan.

**02 PROJECT DOCUMENTS**
- ✓ Change log.
- ✓ Issue log.
- ✓ Lessons learned register.
- ✓ Stakeholder register.

**03 ENTERPRISE ENVIRONMENTAL FACTORS (EEFs).**

**04 ORGANIZATIONAL PROCESS ASSETS (OPA).**

## 13.3 MANAGE STAKEHOLDER ENGAGEMENT    Tools & Techniques 1

**EXPERT JUDGMENT**

**COMMUNICATION SKILLS**
- ✓ The methods of communication identified for each stakeholder in the communications management plan are applied during stakeholder engagement management.
- ✓ The project management team uses feedback to assist in understanding stakeholder reaction to the various project management activities and key decisions.
- ✓ Feedback may be collected in the following ways:
    - Conversations; both formal and informal.
    - Issue identification and discussion.
    - Meetings.
    - Progress reporting.
    - Surveys.

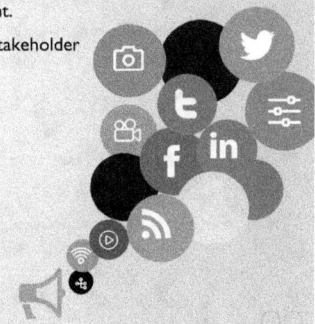

## 13.3 MANAGE STAKEHOLDER ENGAGEMENT — Tools & Techniques 1

### INTERPERSONAL AND TEAM SKILLS

- **Conflict management.** The project manager should ensure that conflicts are resolved in a timely manner.
- **Cultural awareness.** is used to help the project manager and team to communicate effectively by considering cultural differences and the requirements of stakeholders.
- **Negotiation.** is used to achieve support or agreement that supports the work of the project or its outcomes and to resolve conflicts within the team or with other stakeholders.
- **Observation/conversation.** is used to stay in touch with he work and attitudes of project team members and other stakeholders.
- **Political awareness.** is achieved through understanding the power relationships within and around the project.

## 13.3 MANAGE STAKEHOLDER ENGAGEMENT — Tools & Techniques 1

**GROUND RULES**

Defined in the team charter set the expected behavior for project team members, as well as other stakeholders, with regard to stakeholder engagement.

**MEETINGS**

## 13.3 MANAGE STAKEHOLDER ENGAGEMENT

Output

**01 CHANGE REQUESTS**

**02 PROJECT MANAGEMENT PLAN UPDATES**
- ✓ Communications management plan.
- ✓ Stakeholder engagement plan.

**03 PROJECT DOCUMENTS UPDATES**
- ✓ Change log.
- ✓ Issue log.
- ✓ Lessons learned register
- ✓ Stakeholder register.

## 13.4 MONITOR STAKEHOLDER ENGAGEMENT

**MONITOR STAKEHOLDER ENGAGEMENT**

The process of monitoring project stakeholder relationships and tailoring strategies for engaging stakeholders through modification of engagement strategies and plans.

**THE KEY BENEFIT**

It maintains or increases the efficiency and effectiveness of stakeholder engagement activities as the project evolves and its environment changes.

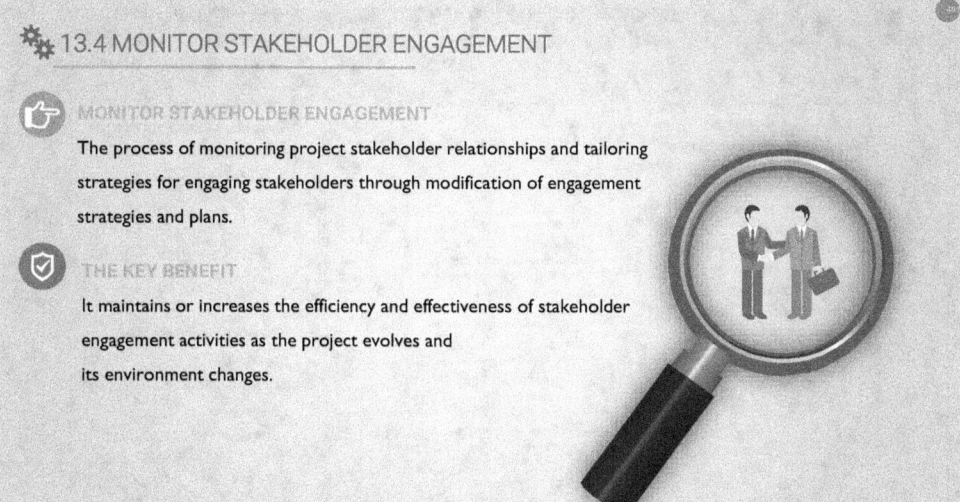

## 13.4 MONITOR STAKEHOLDER ENGAGEMENT

**Inputs**

.1 Project management plan
  • Resource management plan
  • Communications management plan
  • Stakeholder engagement plan
.2 Project documents
  • Issue log
  • Lessons learned register
  • Project communications
  • Risk register
  • Stakeholder register
.3 Work performance data
.4 EEFs
.5 OPA

**Tools & Techniques**

.1 Data analysis
  • Alternatives analysis
  • Root cause analysis
  • Stakeholder analysis
.2 Decision making
  • Multi-criteria decision analysis
  • Voting
.3 Data representation
  • Stakeholder engagement assessment matrix
.4 Communication skills
  • Feedback
  • Presentations
.5 Interpersonal and team skills
  • Active listening
  • Cultural awareness
  • Leadership
  • Networking
  • Political awareness
.6 Meetings

**Outputs**

.1 Work performance information
.2 Change requests
.3 Project management plan updates
  • Resource management plan
  • Communications management plan
  • Stakeholder engagement plan
.4 Project documents updates
  • Issue log
  • Lessons learned register
  • Risk register
  • Stakeholder register

## 13.4 MONITOR STAKEHOLDER ENGAGEMENT — Input

**01 Project management plan**
- ✓ Resource management plan.
- ✓ Communications management plan.
- ✓ Stakeholder engagement plan.

**02 Project documents**
- ✓ Issue log.
- ✓ Lessons learned register.
- ✓ Project communications.
- ✓ Risk register.
- ✓ Stakeholder register.

**03 WORK PERFORMANCE DATA·**

**04 ENTERPRISE ENVIRONMENTAL FACTORS·**

**05 ORGANIZATIONAL PROCESS ASSETS**

## 13.4 MONITOR STAKEHOLDER ENGAGEMENT — Tools & Techniques 1

**01 DATA ANALYSIS**
- Alternatives analysis.
- Root cause analysis.
- Stakeholder analysis.

**02 DECISION MAKING**
- Multi-criteria decision analysis.
- Voting.

**03 DATA REPRESENTATION**
- Stakeholder Engagement Assessment Matrix.

## 13.4 MONITOR STAKEHOLDER ENGAGEMENT — Tools & Techniques 2

**COMMUNICATION SKILLS**
- Feedback. is used to ensure that the information to stakeholders is received and understood.
- Presentations. provide clear information to stakeholders.

**INTERPERSONAL AND TEAM SKILLS**
- Active listening. is used to reduce misunderstandings and other miscommunication.
- Cultural awareness. Cultural awareness and cultural sensitivity help the project manager to plan communications based on the cultural differences and requirements of stakeholders and team members.
- Leadership. Successful stakeholder engagement requires strong leadership skills to communicate the vision and inspire stakeholders to support the work and outcomes of the project.
- Networking. ensures access to information about levels of engagement of stakeholders.
- Political awareness. is used to understand the strategies of the organization, understand who wields power and influence in this arena, and to develop an ability to communicate with these stakeholders.

**MEETINGS**

## 13.4 MONITOR STAKEHOLDER ENGAGEMENT — Output

**01 WORK PERFORMANCE INFORMATION**

**02 CHANGE REQUESTS**

**03 PROJECT MANAGEMENT PLAN UPDATES**
- Resource management plan.
- Communications management plan.
- Stakeholder engagement plan.

**04 PROJECT DOCUMENTS UPDATES**
- Issue log.
- Lessons learned register.
- Risk register.
- Stakeholder register.

HAPPY CUSTOMER

## Have You Any question?

| What | Why | When | Where | How | who |

www.ingramcontent.com/pod-product-compliance
Lightning Source LLC
Chambersburg PA
CBHW052340220526
45465CB00003BA/885